Against Narrow-Mindedness and Chauvinism—Journeying Through Absurd Times with Paul K. Feyerabend

Wolfgang Frindte

Against Narrow-Mindedness and Chauvinism—Journeying Through Absurd Times with Paul K. Feyerabend

Springer

Wolfgang Frindte
Institut für Kommunikationswissenschaft
Friedrich-Schiller-Universität
Jena, Germany

ISBN 978-3-658-50112-9 ISBN 978-3-658-50113-6 (eBook)
https://doi.org/10.1007/978-3-658-50113-6

This book is a translation of the original German edition "Wider die Borniertheit und den Chauvinismus—mit Paul K. Feyerabend durch absurde Zeiten" by Wolfgang Frindte, published by Springer Fachmedien Wiesbaden GmbH in 2024. The translation was done with the help of an artificial intelligence machine translation tool. A subsequent human revision was done primarily in terms of content, so that the book will read stylistically differently from a conventional translation. Springer Nature works continuously to further the development of tools for the production of books and on the related technologies to support the authors.
Translation from the German language edition: "Wider die Borniertheit und den Chauvinismus—mit Paul K. Feyerabend durch absurde Zeiten" by Wolfgang Frindte, © Der/die Herausgeber bzw. der/die Autor(en), exklusiv lizenziert an Springer Fachmedien Wiesbaden GmbH, ein Teil von Springer Nature 2024. Published by Springer Fachmedien Wiesbaden GmbH. All Rights Reserved.

This Springer imprint is published by the registered company Springer Fachmedien Wiesbaden GmbH, part of Springer Nature.
The registered company address is: Abraham-Lincoln-Str. 46, 65189 Wiesbaden, Germany

Preface

> "Anything goes" thus only means: "Do not limit your imagination" (Feyerabend, 1998, p. 163).

On January 13, 2024, Paul Feyerabend would have turned 100 years old.

Anarchist of the philosophy of science, chaotic, enfant terrible, pop star among thinkers, or voodoo priest of epistemology—with such and similar labels, Paul K. Feyerabend was not only described in German cultural sections; his critics from within the scientific community were no less harsh. Some compared him to a guru, and they certainly did not mean that appreciatively. Others saw in him not only the "Salvador Dali of academic philosophy," but "the worst enemy of science." He has also been praised as a positive Dadaist, stimulating provocateur, brilliant philosopher of science, gifted storyteller, singer, actor who enjoyed poking fun at himself and his audience, and—above all—as a committed advocate of scientific pluralism and democratic relativism. In an absurd world, Paul K. Feyerabend developed ideas and sought to bring them to people with theatricality, in order to show ways to confront narrow-mindedness and chauvinism in science and society. There are many signs of the absurdity and ridiculousness of this world. They should not go unmentioned in this book.

I am also a great admirer of Paul K. Feyerabend, for he helped me to free myself from a scientific and ideological rut. I briefly recount this in *Part 1* of this book. In *Parts 2 to 4*, readers will find more detailed information, stories, and conjectures about Paul Feyerabend, his life in Vienna, London, Berkeley, Zurich, and elsewhere, as well as about his work and its impact. I already published a short version of this biographical sketch in 2023 under the title "Against Chauvinism—100 Years of Paul K. Feyerabend" (Frindte, 2023). In *Part 5*, I venture a look at the absurdities of our times, at narrow-minded practices in psychology, at linguistic discrimination, and at chauvinism in crises, disasters, and in the

Russian war of aggression against Ukraine. Paul Feyerabend's ideas serve here as (not always valid) inspiration. Whether he would have liked my excerpts from the Logbook of the Absurd, or would have vehemently rejected my presumption to involve him in these excerpts, he can no longer say. But he is certainly right when he writes: "When one tries to explain one's own ideas, one is forced to provide a 'systematic account' instead of simply telling a story. [...] The difficulty does not lie in the subject matter [...] Is it perhaps the desire to appear grand, profound, and philosophical? But what is more important? To be understood by outsiders or to be regarded as a 'deep thinker'? Writing simply, so that uneducated people can understand, by no means implies superficiality" (Feyerabend, 1995, p. 246). So let's see, for it has not become a simple story. Well, the book is written, and others who may know Paul Feyerabend better will take issue with it—or perhaps not.

A few technical remarks are still permitted: At the end of the book, readers will find a selected index of persons. The selection is a subjective one; included in the index are the names of those whom one should remember—whether in a positive or negative sense—and therefore ought to be able to locate in the text. Paul Feyerabend is not listed in this index. His name appears—well—on many pages of the book, more than 800 times. In the individual chapters, I have also inserted various digressions and additions in the form of highlighted boxes. One may skip over or criticize them. Those familiar with Paul Feyerabend know of his fondness for footnotes. I, too, have been carried away by this fondness and ask for your indulgence. Furthermore, I have, as is my custom, drawn on my own reflections that have already been published elsewhere. I have indicated the relevant passages in the text. This will make it easier for critics to make comparisons. Finally, it should also be noted that quotations from German-language publications have been translated into English by me. This also applies—unless otherwise stated—to the German-language works of Paul Feyerabend.

Incidentally, I also share the opinion: "What we claim to recognize is not independent of us; what is independent of us is and remains unknowable [...] How does that sound? And now I have to get back to my taxes" (Paul Feyerabend to Isaac Ben-Israel on October 25, 1990; Ben-Israel, 2001, p. 100).

November 2023 Wolfgang Frindte
Jena

References

Ben-Israel. (2001). Philosophy and methodology of military intelligence: Correspondence with Paul Feyerabend. *Philosophia, 28*(1–4), 71–101.

Feyerabend, P. K. (1995). Zeitverschwendung. Suhrkamp.

Feyerabend, P. K. (1998, Original: 1996). *Widerstreit und Harmonie*. Trentino Lectures. Passagen Verlag.

Frindte, W. (2023). *Wider den Chauvinismus—100 Jahre Paul Feyerabend.* Wiesbaden: Springer VS.

Acknowledgments

Once again, the collaboration with Springer VS publishing house has proven successful. My special thanks go to Mr. Frank Schindler for the trusting cooperation, as well as to Ms. Madhavi Gopani, Ms. Kalpana Punniyakotti and Mr. Timo Lange for their assistance in completing the final manuscript. I would especially like to thank Dr. Jochen Dreher and Daniel Wilhelm from the Social Science and Philosophy Archive at the University of Konstanz, as well as the staff of the University Library, who enabled me to access the estate of Paul Feyerabend. I would also like to extend my heartfelt thanks to Dr. Grazia Borrini-Feyerabend for her kind encouragement. Grazie mille!

My wife Ina Frindte, with her knowledge as a physicist, not only helped me to better understand some scientific texts. She also subjected the entire book manuscript to her critical scrutiny. I thank her with all my heart and with love.

Contents

Part I
Memories

"All you can do, if you really want to stick to the truth, is *to tell a story,* a story that contains non-repeatable elements side by side with vague analogies to other stories from the same field or from other, distant fields" (Feyerabend, 1992, pp. 152; emphasis in the original).

References

Feyerabend, P. K. (1992; Original: 1989). *Über Erkenntnis. Zwei Dialoge*. Campus.

1 Encouragements

> "*Understanding always comes only after the event* and is hardly ever one of the causes of its occurrence" (Feyerabend, (1986, p. 25; emphasis in original).

I have already commented elsewhere on my path to Paul Feyerabend (Frindte 1998). A brief summary should suffice here: My entanglements with P.F. began in 1981. I was working on the manuscript of my dissertation on game-theoretical modeling of negotiation situations. It was intended to be a critique of the social psychological experiment. At that time, I assumed that social psychological experiments (if at all) only allowed very abstract "insights" into social psychological phenomena and therefore needed to be supplemented and expanded by so-called practical tests on the concrete, manifold, and diverse "object." It seemed to me neither ethically justifiable, hardly theoretically legitimate, nor methodologically grounded to regard the (social-) psychological experiment as the *via regia* of empirical research. And I believed that with this view—which was by no means novel at the time—I could also rely on the method of ascending from the abstract to the concrete, and see in this method the paradigm of a complete research process. "If the true is abstract, then it is untrue. Common sense moves toward the concrete," I had read in Hegel (1979, p. 42; original: between 1805 and 1819), seen confirmed in Marx (Marx, MEW, vol. 13, p. 631), and found again in a note by Lenin (vol. 38, pp. 233). For all the affection I had for experimental research, which seemed to offer me the apparent possibility of bringing clarity to my social psychological thinking, I had been convinced by three years of work as an industrial psychologist after my studies that all scientific theory achieves nothing *unless it grips the masses.*

W. Frindte, *Against Narrow-Mindedness and Chauvinism—Journeying Through Absurd Times with Paul K. Feyerabend*, https://doi.org/10.1007/978-3-658-50113-6_1

In the aforementioned February 1981, H.M., one of my doctoral advisors and himself a critic of lifeless experimental situations, returned my Marxist outpourings on complete research processes with the remark, to the effect: "If you really want to read something good, then get yourself the book 'Against Method' by Feyerabend." H.M. had probably read about this book in a review; he did not own it himself, and my efforts to track it down in a library were unsuccessful. However, I did not try very hard either. I had the system of my dissertation in my head and partly on paper, and did not want to have my circles disturbed by the words of others.

Feierabend—Feyerabend: such a name leaves its traces in memory. Especially since a colleague from philosophy, when I asked who this Feyerabend was, replied: "He's the one with pluralism."

Some problems have something subversive about them. I believe P.F. enjoyed subversion. In 1985, I received the 5th edition of the American version "Against Method" (Feyerabend, 1975) from a West German colleague. I tried to work my way through the English text, initially to improve my English. I eased my struggle with the foreign language by mainly reading the passages that were enlivened by illustrations. Right at the end of the book, in Appendix 5, I came across a passage that almost floored me, because I related it to my own situation. For better understanding, I quote from the German translation of 1986: "A science that claims to possess the only correct method and the only usable results is ideology and must be separated from the state and especially from education. It may be taught, but only to those who have decided to adopt this particular superstition [...] But such special ideologies and skills have no place in *general education,* which prepares the citizen for his role in society. A mature citizen is not someone who has been *instructed* in a particular ideology, such as Puritanism or critical rationalism, and now carries this ideology around like a mental hump, but someone who has learned to form an opinion and then *decided* on what he considers best for himself. He has a certain mental stamina (he does not fall victim to the first ideological street singer he encounters)..." (Feyerabend, 1986, p. 397; emphasis in original).

Little by little, Feyerabend left his mark on my thinking. In the meantime, I was working on my habilitation thesis. "Theory and Experiment in Social Psychology—a Methodological Investigation" was to be its title. I also wanted to engage with various positions in the philosophy of science; for example, with the concept of internal and external validity (e.g., Campbell & Stanley, 1963), those positions in which the concept of falsification in Popper's sense formed the core (e.g., Gadenne, 1976), the methodology of competing research programs by Imre Lakatos (1974), the problem-solving conception of Theo Herrmann (1979),

Stegmüller's non-statement conception (1973), and of course the critical psychology of Holzkamp (e.g., 1973).

Without explicitly addressing Feyerabend in the work, I arrived at conclusions that cautiously distanced themselves from traditional Marxist ideas about the theory-empiricism relationship and perhaps already hinted at some Feyerabendian influence: empirical (experimental) data can neither fully confirm nor fully refute a social psychological theory; a (social-) psychological experiment cannot serve as a full-fledged criterion for the truth of psychological statements; every (social-) psychological research encounters a methodological uncertainty relation, owed to the peculiarities of psychological processes; cooperation between "researchers" and "researched" is a condition for at least not overlooking this uncertainty relation; it is possible to specify procedural rules by which psychological action guidelines can be derived from psychological theories, but those guidelines (or recipes) by which psychologists orient themselves in their practical work are only partially theoretically supported; everyday experiences, implicit theories of psychologists (and their clients), practitioner know-how, the explicit and implicit ethics of psychological work, etc., should not be underestimated as conditions of psychological research (Frindte 1987). Gradually, I bid farewell to the "idea of a universal and stable *method*..." (Feyerabend, 1980, p. 195; emphasis in original).

References

Campbell, D. T., & Stanley, J. C. (1963). Experimental and quasi- experimental designs for research on teaching. In N. L. Gage (Hrsg.), *Handbook of research on teaching* (S. 171–246). Rand McNally.

Feyerabend, P. K. (1975). *Against method. Outline of an anarchistic theory of knowledge*. New Left Books.

Feyerabend, P. K. (1980). *Erkenntnis für freie Menschen*. Suhrkamp.

Feyerabend, P. K. (1986). *Wider den Methodenzwang*. Suhrkamp Taschenbuch.

Feyerabend, P. K. (1992; Original: 1989). *Über Erkenntnis. Zwei Dialoge*. Campus.

Frindte, W. (1998). *Soziale Konstruktionen*. Westdeutscher Verlag.

Gadenne, V. (1976). *Die Gültigkeit psychologischer Untersuchungen*. Stuttgart.

Hegel, G. W. F. (1979). *Werke in zwanzig Bänden, Band 18*, herausgegeben von Eva Moldenhauer und Karl Markus Michel. Suhrkamp.

Herrmann, Th. (1979). *Psychologie als Problem*. Klett-Cotta.

Holzkamp, K. (1973). *Sinnliche Erkenntnis*. Campus.

Lenin, W.I. (1971). *Philosophische Hefte*. Werke, Bd. 38. Dietz.

Marx, K. (1978). Einleitung zur Kritik der Politischen Ökonomie. In *Karl Marx & Friedrich Engels, Werke, Bd. 13*. Dietz.

Stegmüller, W. (1973). Probleme und Resultate der Wissenschaftstheorie und Analytischen Philosophie, Bd. 2: *Theorie und Erfahrung*. Springer.

2 "A scientist [...] is an opportunist"

> "Alone, a person cannot really stand; that is why he likes to join a party, because there, even if not peace, he at least finds reassurance and security" (Goethe, 1972, p. 539; original: 1823).

In the spring of 1988, a gentle dawning began to take place in my mind. The Soviet monthly magazine "Sputnik" had been banned, along with one of the media through which some GDR citizens tried to inform themselves about the course of Soviet perestroika. The general anger over this censorship was great. Psychology students in Jena convinced me to sign a protest letter to the party leadership; other university teachers also took part.

In Berlin, the person responsible for psychology accused me of lacking political clarity. An article on enemy images, a memorandum in which I tried to call for political-psychological research in the GDR, and my refusal to transfer from the Jena Institute of Psychology to a minor subject were the reasons for the criticism. Political psychology was viewed with suspicion by those in charge. What Jena psychology above all needed, they said, was a cognitive psychology that adhered to scientific criteria. Politics and ideology were the business of the party. Timidly, I took note of the criticism, kept my head down, and returned to Jena. My work there on a social-psychological concept, which was to be based on social constructivism and the self-organization paradigm, was meanwhile described by some Jena colleagues as non-Marxist nonsense. My *hour of the comedians.*

Feyerabend (1986, p. 377).

W. Frindte, *Against Narrow-Mindedness and Chauvinism—Journeying Through Absurd Times with Paul K. Feyerabend*,
https://doi.org/10.1007/978-3-658-50113-6_2

My own crisis only reached its peak on October 13, 1989. My friend Jacob, whose joints they had not dislocated and who nevertheless had struggled, said to me that morning, why was I still afraid of the already-dead who thought they could still rule. After my first official and conference trip at the end of September 1989 to the West, to the 15th Congress for Applied Psychology in Munich, I had to report to Berlin again that Friday. Once again, it was about my refusal to leave the Jena Institute and transfer to a minor subject in Leipzig. Psychological peace research was also a reason for criticism.

But neither the topics of conversation nor my supposed insubordination, nor even the extremely rude tone of my interlocutors, are particularly noteworthy—compared to what was happening in this East German country at the time. *To everything there is a season.* Now I had to, and wanted to, do things on my own and in the way I thought appropriate: with the firm resolve never again to let my science be dominated by political interests and currents. "A free society separates state and science" (Feyerabend, 1980, p. 75; original: 1978). It was Paul K. Feyerabend who helped me to breathe and work freely again. Many thanks, dear P.F.

References

Goethe, J. W. (1972, Original: 1823). Aus den Heften „Zur Naturwissenschaft". *Berliner Ausgabe, Bd. 18.* Aufbau Verlag.

Feyerabend, P. K. (1980). *Erkenntnis für freie Menschen.* Suhrkamp.

Feyerabend, P. K. (1986). *Wider den Methodenzwang.* Suhrkamp.

Part II
Waste of Time?

"That is what I wish for: not that my mind should live on, but only love itself" (Feyerabend, 1995, p. 249).

"Waste of Time"—without a question mark[1]: The German autobiography of Paul Feyerabend was published in 1995, one year after his death. He did not waste his time. He enjoyed it. The suffering, the pain, the pleasure and the joy, the theater, singing and writing, the arguments, his opponents, love.

References

Feyerabend, P. K. (1995). *Zeitverschwendung*. Suhrkamp.

[1] The English original edition is titled "Killing Time" (1994).

3 "… The Impression of Reality has Disappeared"

> "Most political and military events made no impression on me" Feyerabend (1995a, p. 62).

Paul Feyerabend was born on January 13, 1924, in Vienna. His mother was from Lower Austria and took her own life in 1943. His father participated as a soldier and later as an officer in the First World War, then worked as an Austrian civil servant in Vienna, and after the annexation of Austria by Nazi Germany became a member of the NSDAP. He died in the late 1950s.

The first apartment Paul Feyerabend can remember, where he lived with his parents, was on Wolfganggasse, in Vienna's 12th district, Meidling, which was then a working-class district and is still shaped by the politics of the SPÖ today. Wolfganggasse is virtually at the center of three addresses where scientists were born or lived for a time, who had a significant influence on Feyerabend's scientific career. In Hofstattgasse in the 18th district, north of Wolfganggasse, Ernst Mach (1838–1916), the great physicist, philosopher, and philosopher of science, lived for a long time. Paul Feyerabend would later call him the most imaginative philosopher of science (Feyerabend, 1987, p. 705). Slightly to the west, in Anton-Langer-Gasse in the 13th district, Karl Raimund Popper was born on July 28, 1902; Feyerabend would later work as his assistant for a time and then make him one of his scientific adversaries.

W. Frindte, *Against Narrow-Mindedness and Chauvinism—Journeying Through Absurd Times with Paul K. Feyerabend*,
https://doi.org/10.1007/978-3-658-50113-6_3

However, Feyerabend would certainly not agree with this brief summary—first he followed Popper, then he broke away from him. In a letter dated June 14, 1985, to his long-time friend, the ethnologist and cultural historian Hans Peter Duerr, Paul Feyerabend mocks such a summary: "That was the sum of my life—he follows a donkey, he stops following him. No wonder I don't want to give interviews or appear on television, because there they always want to examine me in terms of this sum, and no one is interested in my bowel movements, for example, which are much more important to me" (Feyerabend, 1995b, p. 255).

In the distinguished Alleegasse in the 4th district, southwest of Wolfganggasse, Ludwig Wittgenstein was born on April 26, 1889. Certainly, the addresses of these three intellectual giants are several kilometers from Wolfganggasse, but they exemplify the intellectual and cultural density of Vienna's atmosphere in the early 20th century. It is the city of Gustav Klimt, Sigmund Freud, Karl Kraus, Arnold Schönberg, Arthur Schnitzler, Stefan Zweig; it is Red Vienna, where between 1919 and 1933 a social democratic majority governed the city, initiated comprehensive social reforms, and Austromarxism was developed; and it is the city of Austrofascism, with which the Austrian roots of National Socialism are connected.

The apartment on Wolfganggasse was small and sparse: a kitchen, a bedroom-living room, and a study for the father. Here, Paul Feyerabend spent his childhood in a dreamy state until he started school, looking out the window, observing the world, the roadworks, the electrically powered buses, the street artists, listening to the adults in the neighboring apartments arguing and beating their children. At the age of six, Paul started school. After some difficulties with teachers, classmates, and his own restlessness, he apparently managed quite well to cope with the demands of elementary school.

At the age of ten, Feyerabend experienced the February Uprising in Vienna. In 1932, the conservative politician Engelbert Dollfuß became Austrian Federal Chancellor. In 1933, he disempowered the National Council and banned the Communist Party (and a little later the Austrian branch of the NSDAP). The Republican Protection League, a paramilitary organization of the Austrian Social Democratic Workers' Party, was also banned and now operated illegally. When, in February 1934, Austrian police searched for hidden weapons of the Protection League, a bloody uprising ensued. On one side fought members of the Revolutionary

Protection League, on the other side the police and federal army troops. In the fighting, which lasted until February 15, around 1,000 people on the side of the Protection League and more than a hundred on the other side were killed. Numerous Social Democratic officials were subsequently convicted, and eight death sentences were carried out (see also: Österreichische Mediathek, 2023).

Feyerabend writes laconically about his experiences: "The corpses and the blood-spattered streets I saw during the civil war of 1934 in Vienna, and the events of the Nazi era, affected me just as little or, better said, passed by me just as without a trace" (Feyerabend, 1995a, p. 33).

That year, Paul Feyerabend transferred to what was then the Robert Hamerling-Realgymnasium, a school influenced by social democratic educational ideas, which after the "Anschluss" to the "German Reich" was renamed "State Secondary School for Boys," and in which "non-Aryan" students—until their expulsion and extermination—were taught in separate "Jewish classes." At the Gymnasium, Paul Feyerabend learned Latin, French, English, and the natural sciences. "I was a 'top student,' which was indicated in the school reports by a star next to my name" (ibid., p. 38). Later, at about 16 years old, he had the reputation of understanding more about physics and mathematics than his teachers, which did not prevent him from being punished with reprimands. And he read a lot: children's books like Struwwelpeter or the story of Rübezahl, novels by Edgar Wallace, Arthur Conan Doyle, Alexandre Dumas, Jules Verne, Karl May, or Hedwig Courths-Mahler, paperback editions of Goethe, Schiller, Grabbe, Kleist, Ibsen, or Shakespeare. On long walks, he would recite Peer Gynt, Faust, or Shylock. He also came across books by Plato and Descartes, or by the scientists Ernst Mach (1838–1916) and Hugo Dingler (1881–1954), which sparked his interest in philosophy, physics, mathematics, and astronomy. In the school choir, he sang solo parts "[…] and quite impressively, in fact" (Feyerabend, 1995a, p. 48), attended the Vienna Opera and theaters, and—shortly before graduating—took lessons with the baritone and professor of singing Adolf Vogel at the Vienna Academy of Music (1897–1969), who also enjoyed great international success in the late 1930s, including in London, Rome, Buenos Aires, and New York (cf. Österreichisches Musiklexikon online). "My life path was now clearly mapped out: during the day I occupied myself with theoretical astronomy […], in the evening there were rehearsals, singing exercises, and operas […], and at night, finally, astronomical observations" (ibid., p. 53).

A spanner was thrown in the works of the—one might say—highly gifted individual in April 1942, after completing secondary school, by the draft order for the Reich Labor Service.[1] After basic training in Pirmasens, Germany, he was deployed in Brittany. With a tendency toward laziness, a certain defiance, and intensive reading, Feyerabend managed to get through this as well. In December 1943, he was then drafted into the Wehrmacht and a few weeks later volunteered for officer training. He even briefly considered joining the SS. "Because," as he writes, "an SS officer looked better, spoke better, and walked better than an ordinary mortal" (1995a, p. 59). Initially stationed in Yugoslavia, including in Brod, Banja Luka, Novi Sad, Vincovi, and on the so-called Eastern Front against the Soviet Union, he experienced the war as a private, non-commissioned officer, and later as an officer. In March 1944, he was awarded the Iron Cross Second Class because, under enemy fire, he and his soldiers captured a village. In November 1944, at the officer school in Dessau-Rosslau, he gave lectures to officer candidates, among other topics, on the epochs of art history and the Enlightenment. According to his own account, he also spoke about the relationship between Germans and Jews. "For our misfortune," Feyerabend quotes from one of his lectures, "we ourselves were to blame. We must not assign the blame to any Jew, any Frenchman, or any Englishman" (1995a, p. 72). Perhaps he did say that. If so, it would certainly have been considered "treason." Instead, at the end of 1944, he was promoted to lieutenant, reassigned to the "Eastern Front" in Poland, and in January 1945 appointed commander of a bicycle company. After his superiors had retired to the infirmary, Feyerabend became "[...] commander of three tanks, an infantry battalion, and some auxiliary troops from Finland, Poland, and Ukraine" (1995a, p. 73). While fleeing from Soviet troops and—as he writes—out of carelessness, he was wounded by several shots to the hand, face, and back. For him, the war was over. The consequences of these wounds would accompany him for the rest of his life. He would have to live with constant headaches, be unable to walk without a cane, and be impotent.

Feyerabend's autobiography reveals little about his attitudes toward the Nazi regime, the war, or the extermination of the Jews. His parents—like many Austrians—welcomed the "Anschluss" to the German "Reich." Feyerabend was familiar with Hitler's "Mein Kampf" ("The text is unpolished [...] and more a barking than a speech," 1995a, p. 58). He was, however, "moved" by the opening pages of Rosenberg's "Myth of the Twentieth Century." He did notice that his Jewish classmates from school and the Jewish neighbors in Vienna soon disappeared, but he did not give it much thought.

[1] After the annexation of Austria to the Nazi Reich in March 1938, compulsory service in the Reich Labor Service was also introduced for men in Austria on October 1 of that year.

Consider: In 1938, just under 200,000 Jews lived in Austria, about 165,000 in Vienna. After the "Anschluss," pogroms, looting of Jewish shops and apartments, and humiliations of the Jewish population occurred in Vienna and other Austrian cities. From the end of 1938, the "Aryanization" of formerly Jewish businesses took place and—under the significant influence of Adolf Eichmann—the expulsion of Jews from Austria. In February 1941, the deportation of Austrian Jews to ghettos and extermination camps began (Aly et al., 2009). Paul Feyerabend could have known much of this.[2] "During the Nazi era, I paid little attention to the general remarks about Judaism, communism, and the Bolshevik threat. I did not adopt them, but neither did I contradict them" (Feyerabend, 1995a, p. 76).

That in 1942, in the Yugoslav regions where Feyerabend was stationed in 1943, war crimes and pogroms (the Novi Sad massacre in January 1942, the Banja Luka massacre in February 1942) had been committed by Hungarian soldiers or fascist Ustaša troops, he either did not notice or forgot: "We never encountered resistance and hardly thought about the fact that we were an occupying force" (ibid., p. 65).

The German Wehrmacht occupied Yugoslavia in the spring of 1941, supported by Italian, Bulgarian, and Hungarian troops. The Bačka region, which includes Novi Sad, became part of Hungary until the defeat of fascism, and Banja Luka became part of the "Independent State of Croatia." It is quite possible that—in view of the massacres by Hungarian soldiers and Ustaša fascists—the German occupiers were at times and by parts of the (Muslim or ethnic German) population of Yugoslavia regarded as the "lesser evil" (see also Pohl, 2012, pp. 77). Moreover, the Wehrmacht leadership occasionally tried to curb the "senseless Ustaša rampages" in order to legitimize their own acts of violence as necessary and "order-restoring measures" (see also Schmid, 2020, pp. 75).

[2] Stephan Speicher reviewed "Zeitverschwendung," that is, the (German) autobiography from 1995, in the Frankfurter Allgemeine Zeitung and speculated that Feyerabend's apparent forgetfulness might also be related to the circumstances under which the manuscript for the book was written: "The manuscript is the work of a cancer patient; death apparently prevented a more thorough revision" (Speicher, 1995).

Regarding the retreat from the Soviet Union, the reader learns that Feyerabend and his soldiers blew up every house they could find. He did take note of atrocities committed by infantrymen against the "enemy" civilian population, but they did not shock him, "[...] for they were far too strange. But I remembered them, and when I think of them today, I shudder" (Feyerabend, 1995a, p. 67).

The highly detailed accounts of individual episodes from the prewar and wartime periods stand in a curious relationship to the distanced descriptions and explanations of his behavior and experiences during that time. It may be true that much of what happened in those times he only learned about later, after the war, from books or television films, and that what he actually saw, heard, and was involved in hardly affected him. "For me, the German occupation and the war were an inconvenience, not a moral problem, and my actions did not arise from a clear worldview, but from whims and chance circumstances" (ibid., p. 56).

On April 1, 1945, the U.S. Third Army under the command of General George S. Patton crossed the western border of Thuringia. On April 11, the Buchenwald concentration camp near Weimar was liberated.

> On April 7, 1945, the SS began evacuating Buchenwald. At that time, more than 110,000 prisoners were incarcerated in the concentration camp and its 139 subcamps. 56,000 died in the camps. A further 12,000 to 15,000 people lost their lives on the evacuation marches. They starved, died of exhaustion, or were shot by the SS.

On April 16, 1945, the U.S. military assumed governmental authority in Thuringia and on June 9 appointed the former concentration camp inmate and Social Democrat Dr. Hermann Brill as provisional governor. At the beginning of July 1945, U.S. troops left Thuringia, Saxony, and what is now Saxony-Anhalt. The Soviet 8th Guards Army, which had played a key role in the liberation of Berlin, took power in Thuringia.

At this time, Paul Feyerabend was recovering in a military hospital in Apolda, near Weimar. He was on crutches, had his first love affair—during which he became painfully aware of his impotence—and called on the mayor of the city, an anti-fascist, to ask for a job. The mayor—well aware that he had a former Wehrmacht officer before him—assigned him a job in the city's cultural department. There, Feyerabend was now responsible for entertainment and wrote speeches, sketches, and short plays for various occasions (Feyerabend, 1995a, p. 81). Another illness soon brought this episode to an end.

After his recovery, he succeeded in beginning vocal studies in Weimar. The Hochschule für Musik in Weimar had already reopened its doors in July 1945. Well-known music teachers worked as professors at the university during this time: for example, Hermann Abendroth, the famous conductor; Juliane Lerche, a renowned pianist; Erhard Mauersberger, the later Thomaskantor; Hans Pischner, harpsichordist and later cultural politician in the GDR; Maxim Vallentin, who had fled to the Soviet Union in 1935 and after 1945 co-founded the drama department at the university in Weimar; or Josef Maria Hauschild, who was among the most famous vocal teachers in Weimar.

Paul Feyerabend received a scholarship and food ration cards and tested his singing and acting talent, among others, with Hauschild and Valentin, and with the latter he became acquainted with the Stanislavski method.[3] He took lessons, among others, in Italian, harmony, piano, singing, and performance, became a member of the Kulturbund[4], which, as he writes, was the only association of his life.[5] and was a diligent attendee of concerts and theater performances at the German National Theatre in Weimar. Although the main building had been reduced to rubble by a bombing raid in February 1945, performances (until its reopening in 1948) had been relocated to the Weimarhalle.

"I suppose I had a fulfilled life, and yet I was dissatisfied. True to my nature, I did not think about it for long and decided to leave" (Feyerabend, 1995a, p. 84).

It is possible that the new political situation in the Soviet Occupation Zone (SBZ) in general and in Thuringia in particular also contributed to Feyerabend's dissatisfaction. On July 16, 1945, the Provisional Government in Thuringia was dismissed by the Soviet Military Administration in Germany (SMAD) and replaced by a new one, which remained in office until the state elections in autumn 1946. The incumbent president of the government and Social Democrat

[3] A method of acting developed by the Russian/Soviet director and actor Konstantin Sergeyevich Stanislavski, in which actors—unlike what Bertolt Brecht demanded—are supposed to bring their own experiences and feelings into the portrayal of their respective roles. This method has also influenced American actor training as Method Acting (see also: Jansen, 1995).

[4] The "Kulturbund for the Democratic Renewal of Germany" was constituted on August 8, 1945, in Berlin under the chairmanship of Johannes R. Becher. The Thuringian section of the Kulturbund was founded on February 8, 1946 (Zimmer 2019).

[5] That is probably not entirely accurate. From January 1948, Paul Feyerabend was also a paying member of the association of the Österreichisches Colleg, the later Alpbach Forum (Kuby, 2010, p. 1043).

Hermann Brill probably did not fit the image desired by the Soviet occupation authorities. Brill left for Hesse at the end of 1945.

Hermann Brill, who received his doctorate in law from the University of Jena in 1929, was head of the Hessian State Chancellery from 1946 to 1949 and taught as an honorary professor at the University of Frankfurt am Main from 1947 (Overesch, 1992). In addition to various works on law and higher education policy, for example on the place of social sciences in university education (Brill, 1954a), there is a fine article on Karl Kautsky (Brill, 1954b). Why the Social Democrat Brill was not liked by the conservative higher education policymakers in West Germany can be read in an informative article on Wikipedia (Wikipedia.org, 2023).

The SMAD placed—by order—great importance on the rapid restoration of cultural life in the Soviet Occupation Zone (SMAD Orders 50 and 51 from September 1945; see Timofejewa, 2012). Theaters, operas, and museums reopened. However, the tensions between the SED-dominated politics in the SBZ and the sentiments of the population, or the early signs of a cold war, probably did not leave the sensitive, if—by his own account—somewhat politically superficial Paul Feyerabend entirely unaffected.

Perhaps it was simply due to the antifascist plays staged in Weimar, which—from Feyerabend's perspective (Feyerabend, 1995a, p. 84)—did not differ much in their dramaturgy from the plays and dramas of the Nazi era; perhaps there were other reasons as well—in any case, Paul Feyerabend decided to leave Weimar in November 1946 and thus, at the age of 24, to return to Vienna.

References

Aly, G., Heim, S., Herbert, U., Kreikamp, H. D., Möller, H., Pohl, D., & Weber, H. (Hrsg.) (2009). *Die Verfolgung und Ermordung der europäischen Juden durch das nationalsozialistische Deutschland 1933–1945.* Band 2: Deutsches Reich 1938 – August 1939. R. Oldenbourg Verlag.

Brill, H. (1954a). Der Begriff der Sozialwissenschaften: Ein Beitrag zur Auslegung und Anwendung des Bundesbeamtengesetzes. *Zeitschrift für Politik, 1*(1), 86–90.

Brill, H. (1954b). Karl Kautsky: 16. Oktober 1854 – 17. Oktober 1938. *Zeitschrift für Politik, 1*(3), 211–240.

Feyerabend, P. K. (1987). Creativity: A dangerous myth. *Critical Inquiry, 13*(4), 700–711.

Feyerabend, P. K. (1995a). *Zeitverschwendung.* Suhrkamp.

Feyerabend, P. K. (1995b). *Briefe an einen Freund, herausgegeben von Hans Peter Duerr.* Suhrkamp.

Jansen, K. (1995). *Stanislawski – Theaterarbeit nach System. Kritische Studien zu einer Legende.* Lang.

Kuby, D. (2010). Paul Feyerabend in Wien 1946–1955. Das Österreichische College und der Kraft Kreis. In Benedikt, M., Knoll, R., Schwediauer, F., & Zehetner, C. (Hrsg.), Auf der Suche nach authentischem Philosophieren. Philosophie in Österreich 1951–2000. *Verdrängter Humanismus –verzögerte Aufklärung.* Facultas. wuv.

Österreichische Mediathek, 2023. https://www.mediathek.at/akustische-chronik/1919-1938/1934/. Accessed: March, 15, 2023.

Pohl, D. (2012). *Die Herrschaft der Wehrmacht.* Oldenbourg Wissenschaftsverlag.

Schmid, S. (2020). *Deutsche und italienische Besatzung im Unabhängigen Staat Kroatien.* De Gruyter.

Speicher, S. (1995). Disziplin essen Seele auf. Paul Feyerabends Erinnerungen. *Frankfurter Allgemeine Zeitung vom 11. April 1995.*

Timofejewa, N. P. (2012). Deutschland zwischen Vergangenheit und Zukunft: Die Politik der SMAD auf dem Gebiet von Kultur, Wissenschaft und Bildung 1945–1949. In A. O. Tschubarjan & H. Möller (Hrsg.), *Die Politik der Sowjetischen Militäradministration in Deutschland (SMAD): Kultur, Wissenschaft und Bildung 1945–1949* (S. 9–30). K. G. Saur.

Wikipedia.org. (2023). https://de.wikipedia.org/wiki/Hermann_Brill. Accessed: March 15, 2023.

Zimmer, A. (2019a). *Der Kulturbund in der SBZ und in der DDR.* Springer VS.

4 Between Basic Statements, Singing, and Women

> "At space-time location k there is (at least) one non-white swan" (Popper, 2005, p. 78; original: 1934).

As is well known, from 1945 to 1955 Austria was occupied by the Soviet Union, the USA, Great Britain, and France. Vienna, too, was divided into four sectors. The main buildings of the University of Vienna, some of which had been destroyed by bombs, were located in the 1st district of Vienna, which was jointly administered by the occupying powers on a monthly rotating basis. Despite the destruction, teaching resumed as early as May 2, 1945. During the Nazi era, 92 of the 124 full and associate professors were members of the NSDAP (see: History of the University of Vienna a). The Provisional Government of Austria, which was in office between April and December 1945 and under the control of the Allied occupying powers (a coalition of SPÖ, ÖVP, and KPD), enacted various laws for the denazification of the state and society, as a result of which compromised university staff were either immediately dismissed or had to undergo review by special commissions. The effectiveness of the denazification measures is debatable.

In order to enroll at the University of Vienna, Paul Feyerabend also had to undergo a review by an honorary commission. Since he was not a member of the NSDAP, he was classified as unburdened. Initially, in the winter semester of 1946/1947, he enrolled in history, philosophy, and art history, and switched to physics and astronomy in the following summer semester. He attended lectures by, among others, Hans Thirring (1888–1978), Karl Przibram (1878–1973), and Felix Ehrenhaft (1879–1952). Thirring had been dismissed from university service by the Nazis in December 1938 because of his association with the

W. Frindte, *Against Narrow-Mindedness and Chauvinism—Journeying Through Absurd Times with Paul K. Feyerabend*, https://doi.org/10.1007/978-3-658-50113-6_4

("Jewish") theory of relativity and his pacifist stance. He returned to the University of Vienna in 1945, became dean and later vice-dean of the Faculty of Philosophy, and later became involved in the peace movement as an SPD member of the Federal Council. Przibram was also removed from university service in 1938 because of his Jewish heritage, emigrated to Belgium, and there participated in the "Austrian Freedom Front" in resistance against National Socialism. He, too, returned to Vienna in 1946 and took up a professorship at the 2nd Physics Institute of the university in 1947. Ehrenhaft, who, like Thirring and Przibram, was already a recognized scientist before 1933, also lost his position as professor of physics and head of the 3rd Physics Institute at the University of Vienna in 1938 as a Jew. He emigrated to Brazil in 1939 and later to the USA. In March 1947, he too returned to the University of Vienna and was given a visiting professorship there (see: Gedenkbuch Universität Wien).

So it was great scientists whom Paul Feyerabend encountered in Vienna. And they must have impressed him. Thirring was friends with Albert Einstein and Sigmund Freud, founded the Pugwash Peace Conference[1] in 1957 and was twice nominated for the Nobel Peace Prize (Evangelical Museum Austria). Feyerabend appreciated Felix Ehrenhaft for his mischievousness and chutzpah, qualities he himself would have liked to possess at the time.

Among the courses Feyerabend attended in Vienna were lectures by the mathematicians Johann Radon, Edmund Hlawka, and Nikolaus Hofreiter, a former NSDAP member, physics lectures by Theodor Sexl, and astronomy courses by Adalbert Johann Prey, as well as lectures by the psychologist Hubert Rohracher (see also Chap. 12). Feyerabend spent his free time, if it can be called that, at the theater, the opera, and concert halls. He attended discussions on politics and modern science. And he resumed acting and singing lessons.

In 1948, Feyerabend was given an opportunity that would strongly influence his further academic life. In August, he attended the *Alpbach Forum*. "This was the most decisive step of my life" (Feyerabend, 1995, p. 98). The forum was founded in 1945, after liberation from National Socialism, by Otto Molden, then a student in Vienna, and the Innsbruck philosopher Simon Moser as the "Austrian College." Every year since then (to this day), students have met in the Tyrolean mountain village of Alpbach together with representatives from science, politics,

[1] Pugwash is an association—initiated primarily by Bertrand Russell—of international scientists and political experts who strive for a world free of nuclear and other weapons of mass destruction (see also: Kraft & Sachse, 2019).

and the arts to discuss contemporary issues, perform plays, or attend concerts. Of course, there was (and is) also celebration. "Now and then we put on a cabaret. Many affairs blossomed and withered under the moon of Alpbach" (Feyerabend, 1995, p. 98). Three years after the first forum, in August 1948, about 300 to 350 students, professors, and artists came to Alpbach, among them the philosopher of science Karl Raimund Popper.

Popper had earned his doctorate in 1928 under the well-known psychologist Karl Bühler on methodological questions in the psychology of thinking, and subsequently engaged with the Vienna Circle (see below) and logical empiricism. As a result of this engagement, he published one of his major works, "The Logic of Scientific Discovery," in 1934 (Popper, 2005; original: 1934). Knowledge and truth, contrary to the claims of proponents of logical empiricism, cannot be derived or verified by means of formal logic from empirically meaningful singular statements, more or less intersubjectively consistent statements, or observation or basic statements. Rather, it is possible, in short, to refute or falsify a scientific theory by means of empirically meaningful basic statements. With this view, Popper found himself in controversy with the "Vienna Circle" and its leading thinkers, such as Ernst Mach, Ludwig Wittgenstein, or Bertrand Russell, quite early on. Between 1930 and 1935, Popper worked as a secondary school teacher in Vienna. He then spent several months in Great Britain, where he also met Bertrand Russell, the economist Friedrich August von Hayek, and the art historian Ernst Gombrich. In 1937, Popper fled to New Zealand, carrying a letter of recommendation from Russell for a lectureship in philosophy at Canterbury University College in Christchurch. There, he completed the two-volume work "The Open Society and its Enemies." With the support of Hayek and Gombrich, he received a lectureship at the London School of Economics and Political Science in 1946 and was appointed to the chair of logic and scientific theory in 1949 (see in detail: Zimmer, 2019).

And so, Alpbach 1948. After a heated debate about truth, in which Feyerabend also participated, he struck up a conversation with Popper. According to Feyerabend, they talked about music, about Beethoven and Wagner, and perhaps also about basic statements. It may be that Popper was impressed by the young, impetuous Feyerabend. A few years later, Popper would offer him, Feyerabend, a position as a postdoctoral researcher, and Feyerabend would accept it.

In the years after 1948, Feyerabend would return to Alpbach several more times, as a student, lecturer, and seminar leader. He would meet the physicist Erwin Schrödinger (without the cat) and the physicist Lise Meitner, get to know and appreciate the philosophers Rudolf Carnap and Herbert Feigl[2], debate critical rationalism with Joseph Agassi, Hans Albert, or Imre Lakatos. The politician Bruno Kreisky, the writer Arthur Koestler, the actor Qualtinger, and Ernst Bloch also occasionally came to Alpbach.

Alpbach 1948 had consequences. In the Austrian university towns, regional college communities of students formed, who wanted to take up and further discuss the topics of Alpbach in "core circles," "conversations," and "working groups." Based on archival documents, Daniel Kuby (2010) demonstrates that Paul Feyerabend played an active role in this. One of these working groups, in which Feyerabend played a leading role, was the so-called "Kraft Circle," a—according to Feyerabend (1995, p. 104)—"[...] student counterpart to the Vienna Circle." Viktor Kraft (1880–1975) became the academic director of this working group.

The Vienna Circle, founded in 1924 and appearing publicly from 1929 as the "Ernst Mach Association," included philosophers, mathematicians, physicists, and social scientists. Apart from the mathematician Olga Taussky and the physicist Rose Rand, its members were men: the physicist Moritz Schlick, the mathematician and philosopher Hans Hahn, the polymath Otto Neurath, the philosopher and mathematician Rudolf Carnap, the philosopher Herbert Feigl, the mathematician Kurt Gödel, the philosopher and sociologist Edgar Zilsel, and, of course, Victor Kraft. The discussions of the Vienna Circle were occasionally attended by the mathematician Alfred Tarski, the physicist and philosopher Hans Reichenbach, or the philosopher Carl Gustav Hempel. Ludwig Wittgenstein and Karl Bühler were more peripheral to the group. Karl Popper was never a member of the Vienna Circle and, as he writes, never sought an invitation, though he

[2] Feyerabend probably met the philosopher Herbert Feigl (1902–1988) in Vienna in 1954. Feigl had emigrated to the USA in 1930, where he later (1953) founded the Minnesota Center for Philosophy of Science. In 1954, Feigl gave a guest lecture in Vienna (see also: Stadler, 2006, pp. XVI).

would have felt greatly honored had he been invited (Popper, 2012, p. 119). The project that united these luminaries was the effort to address philosophical problems using scientific methods. After the February Uprising of 1934, the "Ernst Mach Association" was banned. In the same year, Hans Hahn, one of the Circle's co-founders, died. In 1936, Moritz Schlick was murdered by a former doctoral student (see: History of the University of Vienna b). In the increasingly anti-democratic, racist, and anti-Semitic climate in Austria, there was no longer any place for these intellectual giants, whose concerns extended beyond the philosophy of science to the democratization of science itself. Most members of the Vienna Circle emigrated in the following years. Viktor Kraft remained in Vienna, but in 1938 was forced into retirement because of his wife's Jewish "descent"; his teaching license was also revoked. After the war, he was rehabilitated, initially worked as a librarian, and in 1947 returned to the Institute of Philosophy at the University of Vienna as an associate professor, becoming a full professor in 1950. It was there that Paul Feyerabend met him.

The "Kraft Circle" was attended by students of philosophy and the natural sciences. They discussed philosophy of science, the theories of relativity, and the world as such. There, Feyerabend met, among others, Elisabeth Anscombe (1919–2001), one of Ludwig Wittgenstein's most renowned students, Ernst Topitsch (1919–2003), who would later become a well-known, though not uncontroversial, sociologist and anti-communist[3] and the Marxist and later friend Walter Hollitscher, who introduced him to the Marxist classics and dialectical materialism.

During this period, it is likely that, through the mediation of Walter Hollitscher, Feyerabend also met Bertolt Brecht, who, after being persecuted by the "House Un-American Activities Committee," had left the USA in 1947 and returned to Europe. Austrian friends wanted to see Brecht at the Vienna

[3] In a later work ("Stalin's War. The Soviet Long-Term Strategy Against the West as Rational Power Politics," 1986, 2nd ed.), Topitsch argued, among other things, that the fascist war of aggression against the Soviet Union was in fact a preventive war against Stalin's attempt to seize world domination. He reaffirmed this view, among other places, in a Festschrift for the Holocaust denier David Irving (Topitsch, 1998a) as well as in a contribution to the "Archiv für Kulturgeschichte" (Topitsch, 1998b).

Burgtheater, and there was even talk of him directing a new Salzburg Festival. However, this failed due to the politics of conservative circles (see also: Mittenzwei, 1986, pp. 265). In any case, Brecht offered Feyerabend a position as an assistant in Berlin. Feyerabend declined and remained in Vienna.

Walter Hollitscher (1911–1986) earned his doctorate in 1934 with Moritz Schlick and Robert Reininger on "Reasons and Causes of the Dispute over the Principle of Causality in Quantum Physics" and subsequently trained in psychoanalysis. As a member of the Communist Party of Austria, he was forced to emigrate in 1938. He went to Great Britain, where he worked as a psychoanalyst until 1945. That same year, he returned to Vienna and initially worked in adult education. With the support of the communist politician Ernst Fischer, Hollitscher received a teaching assignment in 1949, followed by a professorship in logic and epistemology at Humboldt University in Berlin. After publishing his book "Philosophy of Nature" in 1953, he was arrested by the state security service and expelled to Austria. In 1965, the expulsion was revoked and Hollitscher was granted a "regular guest professorship" for philosophical questions of the natural sciences at the University of Leipzig (Pasternack, 2016, p. 178).

It is quite possible that the writer Ingeborg Bachmann also attended the meetings of the "Kraft Circle" from 1948 onwards. In 1950, she completed her doctorate with Viktor Kraft with a dissertation on "The Reception of Martin Heidegger's Philosophy of Existence" (Eberhardt, 2020, p. 5). Joseph McVeigh (2016, p. 49) suggests that Ingeborg Bachmann and Paul Feyerabend were friends and went out together. However, in "Killing Times," or "Zeitverschwendung" Paul Feyerabend does not mention Bachmann.

Both shared a great interest in Ludwig Wittgenstein's philosophy of language. In the spring of 1951, Ingeborg Bachmann tried to meet Wittgenstein in Cambridge, and Paul Feyerabend sought a British Council scholarship to study with Wittgenstein. Both attempts were unsuccessful. Ludwig Wittgenstein died on April 29, 1951. Yet they could not let go of Wittgenstein. Ingeborg Bachmann would later grapple with the conception of language presented by the early Wittgenstein in the "Tractatus Logico-Philosophicus" (e.g., Bachmann, 1993a, b; original 1953). Paul Feyerabend repeatedly returned to Wittgenstein, reviewed the posthumously published "Philosophical Investigations" (Feyerabend, 1955); he would invoke Wittgenstein and reject scientific systems as such (Feyerabend,

1986, p. 380; original: 1975) and develop sympathies for Wittgenstein's pragmatism (Feyerabend, 2005, p. 102).

Two years before Wittgenstein's death, Feyerabend, with the help of Elisabeth Anscombe, managed to invite the philosopher of language, who was then in Vienna, to the "Kraft Circle." He "[...] arrived over an hour late [...] He spoke at length about what one sees when looking through a microscope. Those are the things that really matter, he apparently wanted to say, and not abstract considerations about the relations of 'elementary propositions' and 'theories'" (Feyerabend, 1995, p. 106). With this, if he indeed said it that way, Wittgenstein had struck a chord with Feyerabend. In December 1951, Feyerabend defended his dissertation. It was titled "Zur Theorie der Basissätze" (On the Theory of Basic Statements), (Feyerabend, 1951). The reviewers were Viktor Kraft and the philosopher and literary historian Friedrich Kainz.

As is well known, one does not see abstractions; one can think them, paint them, express them in numbers or sentences, or visualize them in other ways. Love, happiness, and grief as such are abstract words. Scientific theories are also more or less abstract. Wouldn't it be nice, then, if there were very simple and observable facts that, when expressed in a linguistic sentence, designate what is meant by love, happiness, etc., or on which a scientific theory could be based? Wittgenstein had already addressed this problem in the "Tractatus Logico-Philosophicus" (Wittgenstein, 1963; original: 1922), not with love and happiness, but with the relationship between facts, simple sentences, and the world.

Wittgenstein calls (Wittgenstein, 1963, 1.21) elementary propositions the smallest semantic unit with which a fact can be expressed linguistically. Elementary propositions can be either false or true. Everyday sentences, such as those about love, happiness, and grief, or scientific sentences like "The Earth revolves around the Sun," should, according to Wittgenstein, be reducible to such elementary propositions and be testable for truth.

Unfortunately, Wittgenstein does not provide an example of such an elementary proposition. At this point, however, this is just as irrelevant as the scientific turn Wittgenstein later undertook. Much more important is the fact that the leading figures of the Vienna Circle took Wittgenstein's elementary propositions as an occasion to search for criteria by which the meaning of scientific statements could be determined, namely "[...] through logical analysis, more precisely: by reduction to the simplest statements about what is empirically given" (Verein Ernst Mach, 2006, p. 12; original: 1929). The members of the Circle were not always in agreement about the name of these simplest statements, based on sense data. Moritz Schlick and Rudolf Carnap called them fundamental or observation statements, Otto Neurath called them protocol statements. In "The Logic of

Scientific Discovery," Popper speaks of basic statements and not only wants to distinguish himself terminologically, but also, through his principle of falsification, takes controversial substantive positions regarding the linguistic-philosophical movements of the Circle. No singular statement, no observation or basic statement, can confirm the truth of a theory. "There are no pure observations: they are saturated with theories…" (Popper, 2005, p. 89; original: 1934). In his dissertation, Paul Feyerabend adopts the term basic statement, and otherwise also seems to sympathize with Popper's view of the logic of scientific research. According to Feyerabend, basic statements do not differ from other statements in science. They are neither more certain nor less certain than these. Even divergent observations could not refute a basic statement. For that, another theory is needed, "[…] which gives the statement in question its meaning" (Feyerabend, 1951, p. 84). Feyerabend's skepticism toward basic statements, logical empiricism, and the relevance of alternative theories will continue to surface in his later works.[4]

After successfully completing his dissertation in 1951, Feyerabend— as mentioned— applied to the British Council for a scholarship to continue his studies in England. Since Wittgenstein had already passed away, Feyerabend went to Popper at the London School of Economics and Political Science.

Before that, he had made a name for himself not only as a young scholar. He also enjoyed the cultural life in Vienna. From 1947/1948 onwards, he resumed singing lessons, later sang parts from works by Verdi, Puccini, and Bizet, and is said to have had an excellent tenor voice. He attended theater performances, went to the cinema, and loved women. Already during his first stay in Alpbach in August 1948, Paul Feyerabend met Edeltrud (Jaqueline). They married the same year and soon divorced. "I loved Edeltrud, but I was repeatedly distracted by well-dressed and delicately made-up ladies who, at conferences, exuded an aura of beauty, elegance, and wickedness" (Feyerabend, 1995, p. 111). Feyerabend would marry four times in his life; in 1989, his great love, Grazia Borrini.

---

[4] "Today, the term basic statement is rarely used in the philosophy of science; instead, one more often speaks of observation statements. Popper, too, in later works, chose the term observation statement or test statement" (Gadenne, 2019, p. 304).

References

Bachmann, I. (1993a; Original 1953). Sagbares und Unsagbares – Die Philosophie Ludwig Wittgensteins. In *Werke, Band 4, Essays, Reden, vermischte Schriften,* herausgegeben von Christine Koschel, Inge von Weidenbaum & Clemens Münster. Piper.

Bachmann, I. (1993b; Original 1954). Ludwig Wittgenstein – Zu einem Kapitel der jüngsten Philosophiegeschichte. In *Werke, Band 4, Essays, Reden, vermischte Schriften*, herausgegeben von Christine Koschel, Inge von Weidenbaum & Clemens Münster. Piper.

Eberhardt, J. (2020). Existentialphilosophie und Existentialismus. In: M. Albrecht & D. Göttsche (Eds.), *Bachmann-Handbuch*. J. B. Metzler.

Evangelisches Museum Österreich. https://museum.evang.at/persoenlichkeiten/hans-thirring/hans-thirring-physiker-und-pazifist/. Accessed: May 10, 2023.

Feyerabend, P. K. (1951). *Zur Theorie der Basissätze. Dissertation*. Universität.

Feyerabend, P. K. (1955). Wittgenstein's Philosophical investigations. *The Philosophical Review, 64*(3), 449–483.

Feyerabend, P. K. (1986). *Wider den Methodenzwang*. Suhrkamp.

Feyerabend, P. K. (1995). *Zeitverschwendung*. Suhrkamp.

Feyerabend, P. K. (2005). *Die Vernichtung der Vielfalt. Ein Bericht, herausgegeben von Peter Engelmann*. Passagen Verlag.

Gadenne, V. (2019). Karl Poppers Basissätze und Bewährung. In G. Franco (Eds.), *Handbuch Karl Popper* (pp. 303–319). Springer VS.

Gedenkbuch Universität Wien. https://gedenkbuch.univie.ac.at/page. Accessed: 15. March 2023.

History of the University of Vienna a. https://geschichte.univie.ac.at/en/articles/de-nazification-professorate-university-vienna. Accessed: May 10, 2023.

History of the University of Vienna b. https://geschichte.univie.ac.at/de/artikel/der-wiener-kreis. Accessed: May 10, 2023.

Kraft, A., & Sachse, C. (2019). The Pugwash conferences on science and world affairs: Vision, rhetoric, realities. In A. Kraft & C. Sachse (Eds.), *Science, (anti-)communism and diplomacy* (pp. 1–39). Brill Academic Publishers.

Kuby, D. (2010). Paul Feyerabend in Wien 1946–1955. Das Österreichische College und der Kraft Kreis. In Benedikt, M., Knoll, R., Schwediauer, F., & Zehetner, C. (Eds.), *Auf der Suche nach authentischem Philosophieren. Philosophie in Österreich 1951–2000. Verdrängter Humanismus –verzögerte Aufklärung*. (pp. 1041–1056). Facultas. wuv.

McVeigh, J. (2016). *Ingeborg Bachmanns Wien 1946–1953*. Suhrkamp.

Mittenzwei, W. (1986). *Das Leben des Bertolt Brecht, Band II*. Aufbau.

Pasternack, P. (2016). *Die DDR-Gesellschaftswissenschaften post mortem: Ein Vierteljahrhundert Nachleben (1990–2015)*. Berliner Wissenschaftsverlag.

Popper, K. (2005, 11th ed.; Original: 1934). *Logik der Forschung*. Mohr Siebeck.

Popper, K. R. (2012). *Ausgangspunkte. Meine intellektuelle Entwicklung, heraus gegeben von Manfred Lube*. Mohr Siebeck.

Stadler, F. (2006). Paul Feyerabend – Ein Philosoph aus Wien. In F. Stadler & K. R. Fischer (Eds.), *Paul Feyerabend. Ein Philosoph aus Wien*. Springer.

Topitsch, E. (1986). *Stalins Krieg. Die sowjetische Langzeitstrategie gegen den Westen als rationale Machtpolitik* (2nd ed.). Olzog.

Topitsch, E. (1998a). Wider ein Reich der Lüge. In R. Uhle-Wettler (Eds.), *Wagnis Wahrheit.* Arndt.
Topitsch, E. (1998b). Die dunkle Seite des Mondes: Gedanken zur „Schuldfrage". *Archiv für Kulturgeschichte, 80*(2), 453–484.
Verein Ernst Mach. (2006, Original: 1929). *Wiener Kreis. Texte zur wissenschaftlichen Weltauffassung von Rudolf Carnap, Otto Neurath, Moritz Schlick, Philipp Frank, Hans Hahn, Karl Menger, Edgar Zilsel und Gustav Bergmann*, herausgegeben von Michael Stöltzner & Thomas Uebel. Felix Meiner.
Wittgenstein, L. (1963; Original: 1922). *Tractatus logico-philosophicus: Logisch-philosophische Abhandlung*. Suhrkamp.
Zimmer, R. (2019). Karl Poppers intellektuelle Biographie. In G. Franco (Eds.), *Handbuch Karl Popper* (pp. 3–21). Springer VS.

5 London, Vienna, Scientific Perspectives, and a Love in Bristol

"Popper is not a philosopher, he is a pedant—that's why the Germans love him so much" (Feyerabend, 1992, p. 12; original: 1989).

In the autumn of 1952, Feyerabend moved to London for a year of research. He attended lectures by Popper and participated in his seminars. He met Elisabeth Anscombe again, whom he had gotten to know in the "Kraft Circle." Joseph (Joske) Agassi, born in Jerusalem in 1927 and passed away in 2023, later a professor of philosophy in Tel Aviv, Hong Kong, Boston, among other places, became, as Feyerabend writes (1995, p. 131), "almost" his friend. Agassi was a convinced Popperian[1] and also urged Feyerabend to more openly champion Popperian ideas. Feyerabend refused to treat falsificationism like a sacrament and preferred to review Wittgenstein's "Philosophical Investigations," go to the theater, and continue his singing lessons.

In the summer of 1953, Feyerabend's scholarship for his stay at the London School of Economics expired. Popper applied for an extension but was turned down, so Feyerabend returned to Vienna in the summer of 1953. In search of work and income, he wrote several articles and reviews there (e.g., Feyerabend, 1954), a report on scientific life in postwar Austria (Feyerabend, 1955), and translated Popper's "The Open Society and its Enemies" into German (Popper, 2003; original: 1945).

[1] In the 1980s, Agassi—an Israeli citizen—also expressed criticism of Israeli settlement policy and advocated for a stronger separation of state and religion (Agassi, 1984).

W. Frindte, *Against Narrow-Mindedness and Chauvinism—Journeying Through Absurd Times with Paul K. Feyerabend*,
https://doi.org/10.1007/978-3-658-50113-6_5

In the article published in 1954 titled "Physics and Ontology" (Feyerabend, 1954), Feyerabend's interest in topics from Greek antiquity becomes apparent—perhaps for the first time in such detail. The ancient poets and philosophers would accompany him for the rest of his life. Critical Rationalism in general, and falsificationism in particular, would also not let him go, though less out of sympathy, as will be shown.

From 1953 to 1954, Feyerabend took up an assistant position with Arthur Pap (1921–1959), who, on Viktor Kraft's recommendation, had come to the University of Vienna for an academic year as a Fulbright lecturer. Pap was born in Switzerland and initially studied philosophy and literature at the University of Zurich. At the beginning of World War II, he fled to the USA with his Jewish family. There, he continued his philosophy studies at Yale University and Columbia University, earning his doctorate in 1945 under Ernst Cassirer and Ernest Nagel on epistemological problems in physics. In Vienna, Pap gave lectures on analytic epistemology, which he later published as a book with Feyerabend's help (Pap, 1955).

Feyerabend was thus well occupied, occasionally received honoraria for his essays, and he sang again. In between, "[...] he visited some ladies of society when their husbands were not at home" (Feyerabend, 1995, p. 137).

When Popper informed him in 1954 that he had secured the funds to offer Feyerabend an assistant position in London, Feyerabend declined. He did so in a letter to Popper in October 1954, also citing his current singing training and the opportunity to take on a solo part in the opera fragment "Manuel Venegas" by Hugo Wolf at the Vienna Konzerthaus (Kuby, 2010, p. 184). Feyerabend's active participation as a tenor in student concerts at the Vienna Konzerthaus can be documented, for example, on June 23, 1955 (Wiener Konzerthaus, 1955, online).

His interest in philosophy, however, remained. With references from Popper and Erwin Schrödinger, he applied for various positions in Australia, Oxford, and Bristol. He was accepted at the University of Bristol and received a three-year appointment as a lecturer in philosophy of science. "Thus began what is known as my academic career" (Feyerabend, 1995, p. 137). He read works by and about Kant, or about quantum mechanics and its mathematical foundations, as well as detective novels and other things. He wrote reviews, scholarly articles, for example on the quantum theory of measurement (Feyerabend, 1957a), on analytic philosophy (Feyerabend, 1957b), or on the realistic interpretation of experience. In the article "An attempt at a realistic interpretation of experience" (Feyerabend, 1958), which Preston (2002) considers perhaps the most important publication of the "early" Feyerabend, he criticizes the Vienna Circle and positivism, still very much in the spirit of Popper: "[...] the interpretation of an observation-language is determined by the theories which we use to explain what we observe, and it changes as soon as those theories change" (1958, p. 163). The relationships between empiricism, scientific theory, and their justification—those topics that had already been central in his dissertation—thus continued to preoccupy him.

He managed his courses on philosophy of science well, with the support of Joseph Agassi. His lectures on quantum theory, on the other hand, were apparently not particularly captivating at first. However, his talks—for example, on the quantum theory of measurement at the meetings in Alpbach in 1955 and 1956 or at the 9th Symposium of the Colston Research Society in April 1957 in Bristol—by his own account, were apparently better received (Feyerabend, 1995, pp. 147 f.). During this time, Feyerabend also came to know and appreciate the philosophers and physicists Philipp Frank and David Bohm. Frank, among other things, drew Feyerabend's attention—using the example of Galileo's law of inertia—to the fact that theories are often developed in contradiction to rational rules. David Bohm may have had an even greater influence on Feyerabend. He probably met Bohm for the first time at the aforementioned symposium in April 1957 (Kožnjak, 2018).

David Bohm was born in 1917 in the US state of Pennsylvania, studied physics at the California Institute of Technology with Robert Oppenheimer, with whom he received his doctorate in 1943 at the University of California, Berkeley. Later, he worked with Albert Einstein at Princeton University, among others. After refusing to testify before the "House Un-American Activities Committee" against his colleagues, who were accused of communist sympathies, he was initially arrested in 1950 and soon released. He went to Brazil, received a professorship there, and then worked from 1955 to 1957 first as a lecturer and soon as a professor at the Technion in Haifa. From 1957 to 1961, he was a research fellow at the University of Bristol, then a professor at the University of London. David Bohm died in 1992 in London (see also: Olival Freire, 2019).

It may have been Bohm's alternative view of quantum mechanics, as well as the dismissive reactions with which the established community of quantum physicists dogmatically rejected alternative approaches—including Bohm's—that caught Feyerabend's attention. Bohm's argument to never exclude alternative theories, however they may initially appear, ultimately left its mark on the pluralistic views of the later Paul Feyerabend (see also: Van Strien, 2020).

There was certainly no shortage of inspiration in Bristol. And yet Feyerabend seemed to be bored. Apparently, even visits to the Bristol Old Vic theater did little to help. His war injuries also began to trouble him again. He took painkillers in high doses.

And then he met his second wife. Mary O'Neill was one of his students. It must have been a great love and a great celebration when they married in 1956. However, the love and the marriage did not last long. "After I had spent three years in Bristol, I was invited in 1958 to spend a year at the University of California in Berkeley. The invitation came at just the right time. Mary had left me" (Feyerabend, 1995, p. 150).

References

Agassi, J. (1984). *Liberal nationalism for Israel: Towards an Israeli national identity.* Gefen Publishing.

Feyerabend, P. K. (1954). Physik und Ontologie. *Wissenschaft und Weltbild. Monatszeitschrift für alle Gebiete der Forschung, 7*(11–12), 464–476.

Feyerabend, P. K. (1955). *Humanities in Austria: A report on postwar developments.* Library of Congress Reference Department.

Feyerabend, P. K. (1957a). Zur Quantentheorie der Messung. *Zeitschrift für Physik, 148*(5), 551–559.

Feyerabend, P. K. (1957b). Die analytische Philosophie und das Paradox der Analyse. *Kant-Studien, 49,* 238–244.

Feyerabend, P. K. (1958). An attempt at a realistic interpretation of experience. *Proceedings of the Aristotelian Society, 58,* 143–170.

Feyerabend, P. K. (1992; Original: 1989). *Über Erkenntnis. Zwei Dialoge.* Campus.

Feyerabend, P. K. (1995). *Zeitverschwendung.* Suhrkamp.

Kožnjak, B. (2018). The missing history of Bohm's hidden variables theory: The Ninth Symposium of the Colston Research Society, Bristol, 1957. *Studies in History and Philosophy of Science Part B, 62,* 85–97.

Kuby, D. (2010). *„Rational zu sein war damals für uns eine Lebensfrage". Studien zu Paul Feyerabends Wiener Lehrjahren.* Universität Wien.

Olival Freire, J. (2019). *David Bohm. A life dedicated to understanding the quantum world.* Springer Nature Switzerland.

Pap, A. (1955). *Analytische Erkenntnistheorie. Kritische Übersicht über die neueste Entwicklung in USA und England.* Springer Verlag.

Popper, K. R. (2003; Original: 1945). *Die offene Gesellschaft und ihre Feinde. Gesammelte Werke* (Vol. 6). Mohr Siebeck.

Preston, J. (2002). "Paul Feyerabend". The Stanford encyclopedia of philosophy (Summer 2002 Edition). http://plato.stanford.edu/archives/sum2002/entries/feyerabend. Accessed: Jan. 14, 2023.

Van Strien, M. (2020). Pluralism and anarchism in quantum physics: Paul Feyerabend's writings on quantum physics in relation to his general philosophy of science. *Studies in History and Philosophy of Science Part A, 80,* 72–81.

Wiener Konzerthaus – Schülerkonzerte Anton Tausche, 1955. https://konzerthaus.at/konzert/eventid/21429. Accessed: 13. Febr. 2023.

6 Years in Berkeley and Elsewhere

"America was the first country to give me a vague idea of what a culture could be. And by American culture I do not mean Thoreau, Dewey, James, Stevens, or Henry Miller, but variety shows, musicals, wrestling, soap operas: in short, shows and popular entertainment. (Later, in my view, the cultural and racial diversity was added; one of the reasons I did not return to Europe was the uniformity of Europeans.)" (Feyerabend, 1995a, p. 153).

Paul Feyerabend came to Berkeley as a visiting lecturer in September 1958. The invitation was initially limited to one year. Afterwards, the University of California in Berkeley offered him a permanent position. In 1960, he accepted this position and declined both a fellowship and a post at the Minnesota Center for the Philosophy of Science. Feyerabend had already come to know and appreciate the Minnesota Center, founded by Herbert Feigl in 1953, during a sabbatical. He would continue to be a regular guest there, engaging in discussions with Feigl, Grover Maxwell[1], Paul Meehl [2] and several others (Hoyningen-Huene, 1997).

[1] Grover Maxwell (1918–1981) was, at that time, the second-in-command after Herbert Feigl at the Minnesota Center.

[2] Paul E. Meehl (1920–2003), whom Feyerabend would later call an "old friend" in a letter to Hans Peter Duerr from June 1987 (Feyerabend, 1995b, p. 265), worked as a professor of psychology at the University of Minnesota. He was a proponent of critical rationalism and was among the pioneers in applying statistical methods to psychological research (Meehl, 1989); see also *Chap. 12* of this book.

W. Frindte, *Against Narrow-Mindedness and Chauvinism—Journeying Through Absurd Times with Paul K. Feyerabend*,
https://doi.org/10.1007/978-3-658-50113-6_6

The philosopher and philosopher of science Thomas Kuhn (1922–1996) had also been teaching in Berkeley since 1956. When Feyerabend arrived there, Kuhn was working on his major work "The Structure of Scientific Revolutions" (Kuhn, 1967; original: 1962).

As is well known, Thomas Kuhn not only coined terms such as *paradigm* and *paradigm shift* or the distinction between normal science and scientific revolutions. The term *scientific community* also goes back to Kuhn. However, he was inspired in this by Ludwik Fleck. Ludwik Fleck is the author of the book "Entstehung und Entwicklung einer wissenschaftlichen Tatsache" (Genesis and Development of a Scientific Fact), which really ought to be considered a classic of the philosophy of science (Fleck, 1993; original: 1935). However, Fleck's German-language book found little audience in Nazi Germany. After a stay in the Lwów Ghetto and a struggle for survival in Auschwitz and Buchenwald, Fleck died almost forgotten in 1961 in Israel. Only through a brief mention by Thomas S. Kuhn in the preface to "The Structure of Scientific Revolutions" did attentive readers rediscover Fleck's book and his ideas. Kuhn, not least through reading Fleck's book, then famously championed the term *scientific community*, thereby emphasizing the scientific community as a network whose members share more or less congruent scientific perspectives (paradigms). Ludwik Fleck was probably not unknown to Paul Feyerabend either (e.g., Feyerabend, 1987, p. 76).

As Paul Hoyningen-Huene (2002, pp. 62) writes, Feyerabend had already made a name for himself in academic circles at that time as a knowledgeable philosopher of physics. Kuhn, on the other hand, was known only within the narrow circle of historians of science. That Kuhn gained prominence as a philosopher during the 1960s was, in part, due to Feyerabend.

In Berkeley, and later as well, Feyerabend and Kuhn not only argued fiercely with each other but also inspired one another. They were more or less united in their reservations about logical empiricism. However, Feyerabend long rejected Kuhn's concept of *normal science*. "The reason for the aversion to Kuhn's normal science is its dogmatic or quasi-dogmatic element, which Kuhn himself explicitly highlights" (Hoyningen-Huene, 2002, p. 71). In the phase of normal science, a scientific community focuses on an established scientific paradigm, with which the rules for solving scientific problems are prescribed in a relatively dogmatic manner. For Feyerabend, who in the early 1960s was still strongly influenced by critical rationalism in Popper's sense, such dogmatism was abhorrent, as it suppressed critical engagement with alternative theories. On the other hand, with his "[…] almost instinctive aversion to group thinking" (Feyerabend, 1995a, p. 101), the dogmatic or quasi-dogmatic structures in the scientific community were probably not unfamiliar to him.[3] At the very least, until the end of the 1960s, he actively participated in the language games of the Popperians (see also: Hoyningen-Huene, 2002, p. 79).

Thomas Kuhn, who repeatedly modified his concept of paradigm and later spoke of a "disciplinary matrix" (Hoyningen-Huene, 1989, p. 142), describes a paradigm in "The Structure of Scientific Revolutions" as a scientific achievement ("scientific achievements") from which theoretical and methodological standards as well as guiding epistemic values (accuracy, consistency, theoretical simplicity, limits of applicability, etc.) for problem-solving in a scientific community are derived, which are largely binding for the members of the community and provide direction for further research. In doing so, Kuhn also emphasizes the social-normative components in the work of scientific communities—precisely what Paul Feyerabend saw as the dogmatic element in Kuhn's conception.

Feyerabend and Kuhn were not always in agreement about the concept of *incommensurability*. Kuhn introduced it in 1962 in "The Structure of Scientific Revolutions"; Feyerabend also used the term for the first time in 1962 (Feyerabend, 1962).[4] Without going into detail about the differences between the two protagonists' views, the following can perhaps be said: Scientific theories that replace or compete with each other generally do not use the same concepts and are therefore not (or only conditionally) comparable. No concept of one theory can be translated into the concepts of the other theory. According to Feyerabend, *incommensurability* of competing theories is a rare case and actually only occurs when

[3] In the English original of his autobiography, Feyerabend uses the term *group thinking* (Feyerabend, 1994, p. 73). It is not far-fetched to assume that he may have had in mind the term introduced by Irving Janis. Ten years after the Cuban Missile Crisis of 1962, Janis analyzed the meeting minutes of Kennedy's staff and identified group-specific patterns of interaction and communication that pointed to an extreme concentration of communication—group thinking—within the Kennedy staff. This included, among other things, an unrealistic sense of "we," illusions of invulnerability, and a dogmatic suppression of dissenting opinions (Janis, 1972).

[4] In various places, however, Feyerabend mentions that he had already presented his idea of incommensurability in the winter of 1952/53 in a seminar with Popper or in Oxford at the home of Elizabeth Anscombe (Feyerabend, 1980, p. 227; 1986, p. 374).

there are comprehensive competing theories, such as quantum mechanics and classical mechanics or relativity theory and classical physics. Kuhn's concept, by contrast, has a broader range of application (Hoyningen-Huene, 2002, p. 67).

Feyerabend and Kuhn were probably relatively united in the view that a scientific theory can only be refuted by alternative theories and not by empirical observations. Both also considered the history of science to be important. In his work, Paul Feyerabend repeatedly drew on scientifically interesting as well as more or less relevant historical sources and case studies to criticize philosophical conceptions, prescriptions, and rules of science. Thus, he dealt, among others, with the Presocratics, the Sophists, Aristotle or Plato, and Greek myths. He was particularly fascinated by Galileo's methods and arguments (e.g., Feyerabend, 1970a), which will be discussed in more detail later. And Thomas Kuhn describes the development of science as a historical process in which pre-paradigmatic science, normal science, and scientific revolutions alternate. According to Alexander Bird (2008, p. 76), Thomas Kuhn and Paul Feyerabend thus played a significant role in the so-called "Historical Turn in the Philosophy of Science," the shift toward a historical analysis and perspective in the philosophy of science.

So, the times Paul Feyerabend spent in Berkeley during the 1960s were anything but boring. He gave lectures on philosophy of science, held seminars, went to the cinema, attended theater and opera performances, resumed his singing training, discovered his fondness for wrestling shows, married for a third time and got divorced again.

The two souls that seemingly dwelled within him—philosophy of science and singing or acting—soon revealed themselves to be closely connected. For example, he compares the sciences and the arts in order to criticize the increasing separation, specialization, and autonomy of social spheres (Feyerabend, 1967a), or analyzes Eugène Ionesco's "Theater of the Absurd" to show that the boundaries between art and science can be fluid and that both spheres of society must be (ideology-)critical (Feyerabend, 1967b). He would later address the relationship between art as science in greater detail and, after a historical analysis, conclude that sciences are arts, just as arts can be sciences (Feyerabend, 1984a, p. 78).

At the end of the 1960s, Feyerabend, as he writes, was "[…] still a highly sought-after brand name" (1995a, p. 173). He received offers of professorships from London, Berlin, Yale, and Auckland. He accepted the first three. When it became too cold for him in the Northern Hemisphere, he spent several winter semesters between 1972 and 1974 in Auckland, New Zealand. The Georgia Institute of Technology in Atlanta also offered him a position. But even there, it was too cold for him. He became homesick for California and soon resumed his position in Berkeley, which he had initially resigned from, and now commuted between London, Berlin[5], Yale, and Berkeley. This certainly suited his hyperactivity—what else can you call it—very well.

At the London School of Economics, Feyerabend met the mathematician, physicist, and philosopher of science Imre Lakatos.

Imre Lakatos was born on November 9, 1922, in Debrecen, Hungary, as Imre Lipschitz. His mother and grandmother were murdered in Auschwitz. He participated in the resistance against National Socialism. To escape persecution and murder, he changed his name to Imre Molnár during the occupation of Hungary, and after 1945 adopted the name Imre Lakatos. As a communist, he worked from 1949 to 1950 in the Hungarian Ministry of Education, but was then arrested for "revisionist" tendencies and imprisoned for three years. During the Hungarian uprising in the fall of 1956, he fled to Vienna and later to Great Britain. In Cambridge, he earned his doctorate in 1958 on the logic of mathematical discoveries. From 1960 until his death, he taught at the London School of Economics and Political Science, where he was strongly influenced by Popper (Worrall, 1974). However, Lakatos rejected Popper's "naive" falsificationism, i.e., the view that theories must be abandoned if they are falsified, that is, refuted by experimental or other empirical results. According to Lakatos, a theory can only be considered falsified if a better, alternative theory is available. In this, he largely agreed with Feyerabend. Lakatos's sophisticated falsificationism ("sophisticated falsificationism," Lakatos, 1976, p. 182[6]) states that theories are never judged in isolation, but only within the framework of larger research programs, larger theoretical systems. By a research program, Lakatos does not mean an isolated single theory, but a sequence of theories that a) have a conceptually accepted, non-negotiable "hard core" from which one cannot deviate if the research program is to be maintained, and b) a "positive heuristic" by which research problems and hypotheses are defined and divergent findings ("anomalies") are reinterpreted and integrated in terms of the research program (Lakatos, ibid., p. 191). A research program in the sense of Imre Lakatos differs from Kuhn's concept of paradigm in that it has a relatively fixed structure that can be described rationally and compared with other research programs in this respect.

So Feyerabend now found himself at odds with Lakatos. They became friends who could also enjoyably argue about Critical Rationalism.

[5] The appointment to a chair in philosophy of science at the Free University of Berlin took place after a senate meeting in January 1967 (Schwerma, 2015).

[6] Hans Rott (1994, p. 25) also translates "sophisticated falsificationism" as "raffinierter Falsifikationismus."

"Imre and I constantly wrote letters to each other, venting about our affairs, our troubles and worries, and the latest idiocies of our dear colleagues. We differed in character, in our worldview, and in our goals, but we became really good friends. I was shaken and quite upset when I learned of his death" (Feyerabend, 1995a, p. 178).

Imre Lakatos died on February 2, 1974, at the age of 52. In an obituary, Feyerabend honored him as a fascinating personality, an outstanding thinker, and the best philosopher of science of our strange and uncomfortable century (Feyerabend, 1975a, p. 1). Despite all their differences, for example regarding the existence of scientific research programs and their function, Feyerabend and Lakatos agreed on many things, such as the limits of naive falsificationism in Popper's sense or the necessity of studying the history of science to learn how (according to Lakatos) research programs develop or (according to Feyerabend) whether apparent deviations from established theories do not rather point to alternative explanations. Not least, Lakatos and Feyerabend probably also appreciated each other because both were concerned about the increasing irrationality and injustice in this world (Feyerabend, 1975a, p. 2).

London, Berlin, Yale, Berkeley. In Berkeley and Berlin, Feyerabend experienced the political student unrest of the 1960s, the civil rights movements, and the demonstrations against the Vietnam War.

After the university administration tried to ban the activities of the civil rights movements and any political engagement on the Berkeley campus, the *Free Speech Movement* was established there in 1964, in which Feyerabend later also became actively involved. The students demonstrated for the right to free speech and occupied university buildings; there were numerous arrests. Ultimately, the university administration had to give in and lift the bans on political activity on campus. In November 1965, the first major demonstrations against the Vietnam War began on campus. In 1967, Ronald Reagan, the future President of the USA, became Governor of California. Already during his campaign, he had tried to spread stories about the debauched lives of students in Berkeley, to hint at sexual perversions and drug abuse, and promised to clean up the "chaos in Berkeley." One of Reagan's first acts as governor was to dismiss the president of the university. After students occupied a site near the campus in May 1969 to protest the Middle East conflict, Reagan sent over 2,000 National Guard soldiers to Berkeley. Violent clashes with students ensued. One student died, many were injured (see also: Daub, 2022, pp. 138). It should not be forgotten:

A year earlier, on April 4, 1968, Martin Luther King was assassinated, and two months later, on June 6, Robert F. Kennedy also fell victim to an assassination.

During these times, according to his own account (1995a, pp. 167), Feyerabend discussed in his courses works by Daniel Cohn-Bendit, Mao Tse-tung, and Vladimir Ilyich Lenin, debated—in Berkeley—with revolution-minded People of Color about the conflicts between whites and blacks, and invited Vietnamese students to his classes so they could talk about their country and the resistance against the American invasion.

When he was not in London, Berlin, Yale, or Berkeley, he was discussing and giving lectures at conferences in Salzburg, Bellagio, Chicago, Alpbach, and elsewhere. In between, he broke a leg, struggled with a gallstone, had several affairs, suffered heartbreak, and gradually began a scientific turn, so to speak, toward Feyerabend II—a Feyerabend who distanced himself from Popper and Critical Rationalism. In a letter to his friend, the philosopher of science and Critical Rationalist Hans Albert, Feyerabend wrote from London in December 1967: "There are some critical rationalists here in London whose whole lives consist of formulating and refuting hypotheses [...]. To counter this, I am now giving lectures at the London School of Economics, in the heart of Popperland, on the *history* of the sciences, where I show how little Popper's model corresponds to the facts of history [...], and how uninteresting history becomes if one were to rework it according to the model's prescriptions [...]. *Do whatever you want*—I call out to the students—*everything is legitimate* as long as it brings you joy and does not hurt others ..." (Feyerabend, letter to Hans Albert; Baum, 1997, p. 53; emphasis in original). Here he already reveals himself as the Feyerabend who will soon rail against methodological dogmatism.

Hans Albert (1921–2023) studied economics and sociology in Cologne and held the chair of sociology and philosophy of science in Mannheim until his retirement in 1989. He is probably the best-known and most influential representative of Critical Rationalism in the German-speaking world. Albert and Feyerabend met in 1958 at the Alpbach Summer School and, despite significant differences of opinion, became friends. From 1966 to 1973, Albert was "[...] one of Feyerabend's most important correspondents" (Feyerabend, letter to Hans Albert; Baum, 1997, p. 12). In response to the above-quoted letter from Feyerabend, Albert replied on December 15, 1967, among other things: "'Everything is legitimate as long as it brings joy and does not hurt others!' is certainly a very reasonable position, but perhaps still too general to promote progress in knowledge" (Baum, ibid., p. 54).

In 1970, Feyerabend published an article entitled "Against Method: Outline of an Anarchistic Theory of Knowledge." In the conclusions, he writes, among other things: The idea that science should be conducted according to fixed (rational) rules is unrealistic and malicious. It is unrealistic because it presents too simplistic a view of people and their circumstances, and malicious because it restricts people in their possibilities and their humanity (Feyerabend, 1970b, p. 91). Daniel Kuby (2010, p. 180), referring to Clifford Hooker (1972), notes that this work by Feyerabend from 1970 already contains the key elements that would be elaborated five years later in "Against Method" (Feyerabend, 1975b) and would provoke the critics' uproar: the "anarchistic epistemology," the motto "Anything goes," the Lenin and Mao quotations, and the unorthodox interpretation of Galileo.

At the beginning of the 1970s, Paul Feyerabend planned to publish this book together with Imre Lakatos, in which both would debate the pros and cons of rational scientific methods. This did not come to pass. Lakatos died too early, and Feyerabend wrote "Against Method" alone in 1975.

In 1974, the year Imre Lakatos died, Feyerabend took up a visiting lectureship in Brighton for two semesters. John Preston quotes from the memoirs of John Krige, who was among Feyerabend's listeners at that time:

"During the following weeks of that term and for the rest of his year as a visiting lecturer, Feyerabend demolished virtually every traditional academic boundary. He held no idea and no person sacred. With unprecedented energy and enthusiasm he discussed anything from Aristotle to the Azande. How does science differ from witchcraft? Does it provide the only rational way of cognitively organizing our experiences? What should we do if the pursuit of truth cripples our intellect and stunts our individuality? Suddenly, epistemology became an exhilarating area of investigation. Feyerabend created in which people could breathe again. He demanded of philosophers that they be receptive to ideas from the most disparate and apparently far-flung domains, and insited that only in this way could they understand the processes by whereby knowledge grows. His listeners were enthralled, and he held his huge audience until, too ill and too exhausted to continue, he simply began to repeating himself" (Preston, 2002).

Why and how Feyerabend came to Brighton, he seems to have forgotten. He also claims to barely remember his stay in Kassel two years later (Feyerabend, 1995a, p. 208).[7] There, in 1977/1978, he taught two semesters of philosophy of science at the Gesamthochschule. On November 15, 1977, he wrote to his friend Hans Peter Duerr: "[…] I now have a two-year plan: two more years of philosophical nonsense (>commitments<) and then *done*—that is, a switch to the theater" (Feyerabend, 1995b, p. 51; emphasis in original).

But things turned out differently. Erich Jantsch, whom Feyerabend knew from his student days in Vienna, was working at the University of Berkeley at the end of the 1970s and told him that the Swiss Federal Institute of Technology in Zurich was looking for a philosopher of science.

> Erich Jantsch was born in Vienna in 1929 and died in Berkeley in 1980. He advised the *Organisation for Economic Co-operation and Development* (OECD), was one of the founders of the *Club of Rome,* held temporary positions at the Technical University of Hanover, at MIT in Cambridge (MA), and finally as a visiting lecturer in Berkeley. He became well known shortly before his death with the book "The Self-Organizing Universe" (Jantsch, 1979).

Feyerabend applied and, after some delays, got the position in 1980. "Now began ten wonderful years, which I spent partly in Berkeley, partly in Switzerland" (Feyerabend, 1995a, p. 214). In the winter semesters, Feyerabend was in Berkeley, during the summer semesters in Zurich. There he met Paul Hoyningen-Huene, who became a close friend and is still actively engaged with Feyerabend's work today. In Zurich, Feyerabend gave lectures on Plato's Theaetetus and Timaeus as well as on Aristotle and his Physics (see also the lectures by Paul Feyerabend from 1985, recently edited by Michael Hagner and Michael Hampe; Feyerabend, 2023). Handwritten notes and materials by Feyerabend, some of which are kept in the archives of the University of Konstanz, suggest that he must have

[7] In "Erkenntnis für freie Menschen" ("Science in a Free Society"), however, he reconstructs in great detail a discussion that must have taken place after a lecture on January 18, 1978, at the Gesamthochschule Kassel (Feyerabend 1980, pp. 283 ff.), and the introduction to the selected writings "Der wissenschaftliche Realismus und die Autorität der Wissenschaften" (Scientific Realism and the Authority of the Sciences) is dated "Kassel, December 1977" (Feyerabend, 1978, p. 3).

prepared thoroughly for these lectures (Paul Feyerabend Archive, University of Konstanz). He opened his seminars to the general public; he invited followers of Rudolf Steiner, asked Friedrich Dürrenmatt to give a lecture, and organized panel discussions dealing with "borderline problems of the sciences," including a seminar on parapsychology (see also: Feyerabend, 1984b and *Ch. 12*).

In 1983, Paul Feyerabend met and fell in love with Grazia Borrini in Berkeley. They married in January 1989. A year later, he retired in Berkeley, and in 1991 he retired in Zurich as well.

Elmar Holenstein, Professor of Philosophy at the Swiss Federal Institute of Technology in Zurich from 1990 to 2002, writes about Feyerabend's time in Zurich, among other things: "At ETH he (Feyerabend, WF) succeeded, as only few have, in encouraging both students and his colleagues, specialists and laypeople alike, to reflect on philosophical questions. Here, he was taken seriously, even if one did not share his philosophy. Why? When he was appointed to Zurich, he invoked the Enlightenment philosopher Kant. In Feyerabend himself, one found the combination of two qualities that philosophers of the 18th century held as their ideal: expertise and benevolence. [...] And his advocacy for the powerless of this earth, the socially and ideologically disadvantaged and marginalized, was highly valued" (Holenstein, 1994, p. 302).

Grazia Borrini-Feyerabend was born in Italy in 1952. She studied public health and physics in Berkeley and Florence and worked for many years in advisory and administrative bodies to promote nature conservation on an international scale. In 2008, she was among the co-founders of the ICCA Consortium (International Congress and Convention Association, a non-profit organization that sees itself as a hub of the international conference industry). Until 2019, she served as coordinator of the ICCA Consortium. Today, she heads the *Paul K. Feyerabend Foundation,* which aims to support projects, organizations, and individuals committed to solidarity with communities in need (https://www.pkfeyerabend.org/en/).

"I no longer wanted to write essays, just finish a short book, occasionally give lectures, and use the fees for trips with Grazia. I thought I could now read, walk in the woods, and take care of my wife. But it didn't turn out that way" (Feyerabend, 1995a, p. 241).

Paul K. Feyerabend died on February 11, 1994, at the age of 70 in Switzerland from a brain tumor. He found his final resting place in a grave of honor at the Vienna Central Cemetery (Südwestfriedhof).

References

Baum, W. (Hrsg.). (1997). *Paul Feyerabend – Hans Albert. Briefwechsel.* Fischer Verlag.

Bird, A. (2008). The historical turn in the philosophy of science. In S. Psillos & M. Curd (Eds.), *Routledge companion to the philosophy of science* (S. 67–77). Routledge.

Daub, A. (2022). *Cancel culture transfer.* Suhrkamp.

Feyerabend, P. K. (1962). Explanation, reduction, and empiricism. In H. Feigl & G. Maxwell (Eds.), *Minnesota studies in the philosophy of science,* Bd. 3. University of Minnesota Press.

Feyerabend, P. K. (1967a). On the improvement of the sciences and the arts, and the possible identity of the two. In R. S. Cohen & M. W. Wartofky (Eds.), *Proceedings of the Boston Colloquium for the philosophy of science 1964/1966.* D. Reidel Publishing.

Feyerabend, P. K. (1967b). Theater als Ideologiekritik. Bemerkungen zu Ionesco. In E. Oldemeyer (Hrsg.), *Die Philosophie und die Wissenschaften. Simon Moser zum 65. Geburtstag.* Verlag Anton Hain.

Feyerabend, P. K. (1970a). In defence of classical physics. *Studies in History and Philosophy of Science, 1*(1), 59–85.

Feyerabend, P. K. (1970b). *Against method. Outline of an anarchistic theory of knowledge.* University of Minnesota Press. https://conservancy.umn.edu/handle/11299/184649. Accessed: March 5, 2023.

Feyerabend, P. K. (1975a). Imre Lakatos. *The British Journal for the Philosophy of Science, 26*(1), 1–18.

Feyerabend, P. K. (1975b). *Against method. Outline of an anarchistic theory of knowledge.* New Left Books.

Feyerabend, P. K. (1978). *Der wissenschaftstheoretische Realismus und die Autorität der Wissenschaften.* Vieweg & Sohn.

Feyerabend, P. K. (1980). *Erkenntnis für freie Menschen.* Suhrkamp.

Feyerabend, P. K. (1984a). *Wissenschaft als Kunst.* Suhrkamp.

Feyerabend, P. K. (1984b). Was heißt das, wissenschaftlich zu sein? In P. Feyerabend & C. Thomas (Hrsg.), *Grenzprobleme der Wissenschaften* (S. 385–397). Eidgenössische Technische Hochschule, Verlag der Fachvereine.

Feyerabend, P. K. (1986). *Wider den Methodenzwang.* Suhrkamp.

Feyerabend, P. (1987). Putnam on incommensurability. *The British Journal for the Philosophy of Science, 38*(1), 75–81.

Feyerabend, P. K. (1992; Original: 1989). *Über Erkenntnis. Zwei Dialoge.* Campus.

Feyerabend, P. K. (1994). *Killing time.* The University Press of Chicago Press.

Feyerabend, P. K. (1995a). *Zeitverschwendung*. Suhrkamp.

Feyerabend, P. K. (1995b). *Briefe an einen Freund, herausgegeben von Hans Peter Duerr.* Suhrkamp.

Feyerabend, P. K. (2023). *Historische Wurzeln moderner Probleme.* Herausgeber: Michael Hagner & Michael Hampe. Suhrkamp.

Fleck, L. (1993; Original: 1935). *Entstehung und Entwicklung einer wissenschaftlichen Tatsache.* Suhrkamp.

Holenstein, E. (1994). Zum Tod von Paul Feyerabend. 1924–1994. *Zeitschrift für philosophische Forschung, 48*(2), 300–302.

Hooker, C. A. (1972). Critical Notice of M. Radner & S.Winokur (Eds.). Analyses of theories and methods of physics and psychology; Minnesota Studies in the Philosophy of Science. *Canadian Journal of Philosophy, 1*(3–4), 393–407; 489–509.

Hoyningen-Huene, P. (1989). *Die Wissenschaftsphilosophie Thomas S. Kuhns.* Vieweg & Sohn.

Hoyningen-Huene, P. (1997). Paul Feyerabend. *Journal for General Philosophy of Science, 28*(1), 1–18.

Hoyningen-Huene, P. (2002). Paul Feyerabend und Thomas Kuhn. *Journal for General Philosophy of Science, 33*(1), 61–83.

Janis, I. (1972). *Victims of groupthink: A psychological study of foreign policy decisions and fiascoes.* Houghton Mifflin.

Jantsch, E. (1979). *The self-organizing universe: Scientific and human implications of the emerging paradigm of evolutio*n. Pergamon Press.

Kuby, D. (2010). *„Rational zu sein war damals für uns eine Lebensfrage". Studien zu Paul Feyerabends Wiener Lehrjahren. Diplomarbeit.* Universität Wien.

Kuhn, T. (1967, Original: 1962). *Die Struktur wissenschaftlicher Revolutionen.* Suhrkamp.

Lakatos, I. (1976). Falsification and the Methodology of Scientific Research Programmes. In S. G. Harding (Eds.), *Can theories be refuted?* (S. 205–259). D. Reidel Publishing Company. https://link.springer.com/chapter/https://doi.org/10.1007/978-94-010-1863-0_14. Accessed: Febr. 12, 2023.

Meehl, P. E. (1989). Paul E. Meehl. In G. Lindzey (Hrsg.), *A history of psychology in autobiography* (S. 336–389). Stanford University Press.

Paul Feyerabend Archiv. Universität Konstanz. https://www.kim.uni-konstanz.de/phil-archiv/bestaende/paul-feyerabend/. Accessed: Febr. 7, 2023.

Preston, J. (2002). "Paul Feyerabend". The Stanford Encyclopedia of Philosophy (Summer 2002 Edition). http://plato.stanford.edu/archives/sum2002/entries/feyerabend. Accessed: Jan. 14, 2023.

Rott, H. (1994). Zur Wissenschaftsphilosophie von Imre Lakatos. *Philosophia naturalis, 31,* 25–62.

Schwerma, J. (2015). Universitätsarchiv der Freien Universität Berlin. Protokolle des Akademischen Senats 1948–1969. https://www.fu-berlin.de/sites/uniarchiv/ressourcen_pdf_/Findbuch_AS1_2015.pdf. Accessed: Febr. 5, 2023.

Worrall, J. (1974). Nachruf auf Imre Lakatos: Imre Lakatos (1922–1974): Philosopher of Mathematics and Philosopher of Science. *Journal for General Philosophy of Science, 5*(2), 211–217.

Part III
Key Works—Selection

"I have forgotten many things, and I have mixed up others" (Feyerabend, 1992, p. 180).

Even though he claimed that a large part of his life had unfortunately been spent idly loitering and waiting (Feyerabend, 1992, p. 181), Paul Feyerabend was a diligent reader, witty storyteller, and an intellectually vibrant as well as productive author. Eric Oberheim, with the help of Grazia Borrini-Feyerabend, undertook the effort to compile an overview of Feyerabend's publications up to 1997 (Oberheim, 1997). In the following chapters, I will limit myself to four books that, for me, are the key to understanding Feyerabend's work. These are the German translations of "Against Method" (1975; German edition: 1986), "Science in a Free Society" (1978; German edition: 1980), as well as the posthumously published books "Philosophy of Nature" (2009) and "Conquest of Abundance" (1999; German edition: 2005). Where appropriate, references will also be made to other publications by Paul Feyerabend—in the spirit of "Somehow it'll work out."

References

Feyerabend, P.K. (1992; Original: 1989). Über Erkenntnis. Zwei Dialoge. Campus.

7 Against the Tyranny of Method

"So, my next book is an ode to absurdity" (Paul Feyerabend 1974 in a letter to Hans Peter Duerr; Feyerabend, 1995b, p. 25).

In 1975, Feyerabend published his book "Against Method" (Feyerabend, 1975); in 1976, it appeared in German under the title "Wider den Methodenzwang".[1] As he writes, it is a collage of descriptions, analyses, and arguments, some of which he had already expressed and defended in earlier works (Feyerabend, 1995a, p. 189). With this book, Feyerabend established himself as an impetuous critic of rationalism, an excellent connoisseur of the history of science, and a rebel who is averse to any glorification of science. The book contains that slogan which has since seemed inseparably linked to Feyerabend's views: "Anything Goes" (Feyerabend, 1986, p. 32). Depending on one's taste, sympathy with Paul Feyerabend, or ideological orientation, the slogan can be translated as "Do what you want," "Everything is permitted," or "Something always works." For critics, "Anything goes" is the anarchistic principle that, from Feyerabend's perspective, should underlie all scientific activity and the democratic constitution of society.[2]

---

[1] Feyerabend revised the English original as well as the German translations from 1976 and 1983 several times, so that today at least six versions of "Wider den Methodenzwang" exist, along with numerous translations into other languages, such as Chinese, French, Italian, Japanese, Dutch, Spanish, and Turkish (see also Oberheim, 1997). I rely on the 1986 German edition, which is identical to the 1983 version.

[2] The Marxist philosopher Herbert Hörz, who counts Paul Feyerabend among his friends (Hörz, 2019, p. 292), apparently also misunderstood "Anything Goes" (Hörz, 1983, p. 790).

W. Frindte, *Against Narrow-Mindedness and Chauvinism—Journeying Through Absurd Times with Paul K. Feyerabend*, https://doi.org/10.1007/978-3-658-50113-6_7

Feyerabend himself suspects that most critics apparently stopped reading the book right after the first appearance of "Anything goes." The subsequent case studies were "either too difficult or too detailed for them, or, taking the emptiness of their own heads as a model, they already took the empty and unexplained principle for the thing itself" (Feyerabend, 1986, p. 383). The case studies in question are mainly historical examples, such as Galileo Galilei's scientific efforts to defend the Copernican worldview against the Ptolemaic one.

As is well known, in the geocentric worldview developed by the Greeks and fully elaborated by Claudius Ptolemy (ca. 100—ca. 170 AD), the earth rests at the center of the world and does not move. The (then known) planets, the sun, and the moon move around the earth. Nicolaus Copernicus (1473–1543) put an end to this and, with the heliocentric model, initiated a new beginning, also known as the Copernican Revolution: a) The earth rotates on its own axis and b) orbits the sun, as do the other planets. Supporters of the Ptolemaic worldview, whom Feyerabend calls Aristotelians, vehemently opposed both assumptions.[3] Paul Feyerabend illustrates this using assumption a) as an example. The Aristotelians used the so-called tower argument to reject this assumption: If the earth rotated daily, a tower from which a stone is dropped would be carried along by the earth's rotation. During the time it takes the stone to fall, the tower would move with the earth's rotation, and the stone would therefore land many hundreds of meters away from the tower (Feyerabend, 1986, p. 90). In reality, however, the stone falls straight down. Therefore, the earth cannot have moved. Thus, the Copernican view of the earth's motion seemed refuted. Galilei, Feyerabend continues, admits that the sensory content of the observation is correct, that is, one sees the stone fall vertically, and yet he tried to refute the tower argument.

In doing so, Galilei worked with "rational arguments" and a "psychological trick" (Feyerabend, 1986, p. 105). Among the rational arguments is Galilei's call

[3]As a reminder: On February 17, 1600, Giordano Bruno (1548–1600) was burned at the stake in Rome for heresy. He had banished the earth from the center of the universe, was the first to emphasize the infinity and eternity of the world, thereby dethroning the Christian God and laying the philosophical foundations for a scientific worldview. With this view of the world, Johannes Kepler (1571–1630) sought the laws governing planetary orbits. He published his observations and laws in 1609 and 1619. A few years later, Galileo Galilei followed with his 1632 work "Dialogo di Galileo Galilei sopra i due Massimi Sistemi del Mondo Tolemaico e Copernicano" (short: Dialogo), in which he discusses the Ptolemaic and Copernican world systems, considers the latter more accurate, and thus fell into the clutches of the Inquisition, from which, as is well known, he could only escape by submission.

to see what one observes in a new light and to describe it in a new language. In this language, alongside the principle of inertia, there also appears the aspect we now know as the principle of relativity. If, as Galilei does, one assumes inertia—that the earth (and thus the tower as the observation point) moves eastward together with the falling stone—then, due to the co-motion of the observation point, the fall of the stone is perceived as a straight line. The principle of relativity described by Galilei states that "our senses perceive only relative motions and are blind to motions in which objects participate to the same degree" (1986, p. 115). Thus, an observer on the tower can perceive the falling stone, but not immediately the motion of the earth as well.

In any case, according to Feyerabend, these are speculative ad hoc hypotheses introduced by Galilei, which hardly or not at all correspond to sensory perception and are presented with a considerable degree of persuasiveness. Galilei does not provide relevant arguments for introducing these ad hoc hypotheses, but instead refers to "what everyone supposedly already knows" (1986, p. 117). This is the psychological trick, which can hardly be explained by the principles of critical rationalism. Galilei succeeded because, in his scientific approach, he violated important rules (or principles) of the scientific method, invented by Aristotle and later canonized by the logical positivists (such as Carnap and Popper) (ibid., p. 169). Such rules as clarity, precision, objectivity, consistency, or truth are ultimately unsuitable if one intends to turn to the rich historical material without diluting it or satisfying one's craving for intellectual security. Galilei succeeded "because he did not follow these rules…" (Feyerabend, 1986, p. 169).

For Feyerabend, Galilei's way of working, exemplified by further examples—such as his lunar observations—is virtually the paradigm of how knowledge works. "The case of Galilei's lunar observations is only a small part of my argument that Galilei did not proceed 'scientifically' and could not have made his discoveries 'in a scientific way'" (1986, p. 183).

The conclusion Feyerabend tries to draw from the case studies can perhaps be summarized as follows, to return once more to "Anything goes": "It is therefore clear that the idea of a fixed method or a fixed theory of rationality is based on an all too naive view of human beings and their social relations. Anyone who turns to the rich material provided by history and does not intend to dilute it in order to satisfy their base instincts, namely the craving for intellectual security in the form of clarity, precision, 'objectivity', 'truth', will realize that there is only *one* principle that can be defended under *all* circumstances and at *all* stages of human development. It is the principle: *Anything goes*" (1986, p. 31; emphasis in the original).

With "Anything goes," Feyerabend, as he notes elsewhere, does not want to introduce a new *principle*. Rather, it is "a somewhat humorous depiction of the rationalist's situation: he wants to have general principles, but in view of the material I provide, he must empty them more and more of all content. 'Anything goes' is all that remains" (Feyerabend, 1980, p. 101; original: 1978a).

Whether he actually intended thereby to introduce an "anarchist" or "Dadaist epistemology" (Feyerabend, 1977), or whether this too is rather a humorous term to describe the practice of scientific activity, shall be left open for now.

Paul Feyerabend recounts in various places a meeting with Carl Friedrich von Weizsäcker in Hamburg. It apparently took place in 1965 (Feyerabend, 1976, p. 384; Jung, 2000; p. 162), although Feyerabend mentions 1968 elsewhere (Feyerabend, 1978b, p. 203). In the discussion with Weizsäcker, Feyerabend realized that general methodological reasons only hinder concrete research rather than promote it. "Professor von Weizsäcker," Feyerabend later said, "is therefore largely responsible for my transition to 'anarchism'—but he was by no means pleased when I told him this in 1977" (Feyerabend, 1980, p. 231).

In a letter to Hans Albert dated August 5, 1969, Feyerabend confesses, among other things, his growing interest in Dadaism, which in his view differs from anarchism in that Dadaism has more humor, "[…] that a Dadaist would certainly never kill a person, and that he presents his works to the public with humorous and critical intent" (Feyerabend, letter to Hans Albert; Baum, 1997, p. 121).

At the very least, Feyerabend sympathizes with an "epistemological anarchism," which he wishes to distinguish from political and religious anarchism. The epistemological anarchist—unlike the political and religious anarchist—is not a skeptic, but does not shy away from "[…] defending the most trivial or the most outrageous statement […] for to him there is no eternal loyalty and no eternal aversion to any institution or ideology […]. No view is so 'absurd' or 'immoral' that he would not be willing to consider it or act upon it, and no method is indispensable to him. The only thing he unequivocally and unconditionally opposes are general principles, general laws, general ideas such as 'truth,' 'justice,' 'love,' and the behavior they engender, even though he does not deny that it is often tactically correct to act as if such laws (principles, ideas) existed and as if he believed in them" (1986, p. 249).

Thus, the epistemological anarchist is an opportunist and a pragmatist, convinced that there are many ways to approach nature and society, as well as many possibilities for judging these approaches. However, there are no objective criteria for this.

It is easy to imagine that many of Feyerabend's critics have grappled with such a view, and this will be mentioned later. But first, attention should be drawn to another aspect: Feyerabend's hints that some manifestations of political anarchism may not have been entirely unsympathetic to him. "Against Method" was written during the time of student unrest in the USA and Europe. Daniel Cohn-Bendit, the theoretically versed, rhetorically gifted, and action-oriented spokesman of the student unrest in France, described himself in a SPIEGEL interview in May 1968 as an "anarchist Marxist." "For me," says Cohn-Bendit, "Marx's fundamental analysis is correct, the analysis of capitalist society. But I completely reject the organizational forms adopted by the communist movement." They do not bring about a new society, but only authoritarian rule (Der Spiegel, 26.05.1968).[4] That could just as well be a statement by Feyerabend. As mentioned, he knew Cohn-Bendit, had read and even cited Marx, Lenin, Trotsky, and Mao Tse-tung.

A somewhat cryptic example can be found in "Against Method" (1986, p. 194). In order, as he writes, to explain that a "phase shift" between progressive ideas (e.g., the movement of the Earth) and regressive means and theories of observation (e.g., the Ptolemaic worldview) cannot be resolved by logical procedure, he cites in footnotes, among others, Marx's statements about *"the unequal relationship of the development of material production, e.g., to artistic production"* from the 1857 introduction to the "Grundrisse of the Critique of Political Economy," which was not published during Marx's lifetime (see Marx, 1983, p. 15). The quotation is interesting if one wishes to consider the dialectic of social base and cultural superstructure, and, for example, to reflect on an autonomous art that exists and develops relatively independently of material production, and cannot be seen—according to a naïve materialist reflection theory—as a mere mirror of production. Certainly, such a dialectic is ultimately bound to fail under purely logical analysis. After all, these are contradictory, historical processes that seek their own forms of movement. Such processes interest Paul Feyerabend.

[4]At the time of the interview, Daniel Cohn-Bendit was in Berlin. On May 21, 1968, he spoke during a rally of the Socialist German Student Union in the Auditorium Maximum of the Free University of Berlin. Afterwards, the French government denied him re-entry to France.

However, the Marx quotation does not seem entirely appropriate to the mentioned phase shift between new theories and old ideas. The same is likely true for Feyerabend's references to Trotsky (on the non-simultaneity of different aspects in the historical process) or to Lenin's thesis of the "uneven development of capitalism."[5] Perhaps Feyerabend's references to Marx, Lenin, or Trotsky are also due to the revolutionary fashions of the time, an era in which students, as on February 18, 1968 in West Berlin, demonstrated against the Vietnam War with portraits of Che Guevara, Lenin, Rosa Luxemburg, or Hồ Chí Minh, or argued with quotations from Lenin, Marx, or Trotsky. In June 1970, Feyerabend wrote in a letter to Hans Albert, among other things, that in Yale he had met verbose critics of reason, against whom the best remedy was a small dose of Lenin, for example *What Is to Be Done?* or *Left-Wing Communism* and the like. "[…] and I must say, what is happening *today* on the left is cellar-deep (as opposed to 'towering high') below the level of Lenin, Marx, and Trotsky" (Feyerabend, letter to Hans Albert; Baum, 1997, p. 223; emphasis in the original).

Feyerabend's sympathy for Daniel Cohn-Bendit is likely more comprehensible than his quotations from Marx, Lenin, or Trotsky. Whether Feyerabend saw, heard, or even met Dany le Rouge in Berlin is uncertain, but conceivable. In February 1968 and again in May of the same year, both were in West Berlin.

On November 18, 1968, Paul Feyerabend wrote to Imre Lakatos, among other things: "I have finished Cohn-Bendit, and I am wholly on his side. He is against theories; so am I. He is against organisations; so am I. He is against 'leaders' be they professors who 'know' or generals who command; so am I. He is for joy and against sacrifice; so am I […] I now see my Against Method as a weak and stumbling prologue to what others have done much better: Cohn-Bendit, for example" (quoted in Martin, 2019, p. 23).

The epistemological anarchist Feyerabend, who not only advocates the separation of church and state but also supports the separation of state and science (Feyerabend, 1986, p. 385), thus indeed sympathized somewhat with political anarchism, but resists any suggestion that he endorses violence.

With "Against Method," Paul Feyerabend caused an uproar in circles where scientists saw themselves as guardians of the only truth, as experts in knowledge, as champions of pure method—yet ultimately, they are only human: loud, cheeky,

[5] Here, Feyerabend refers, among other things, to Trotsky's speech "The School of Revolutionary Strategy," which he delivered in 1921 in Moscow at a meeting of the Moscow Party Organization (Trotzki, 2016; original: 1921), and to Lenin's essay "Das rückständige Europa und das fortgeschrittene Asien" (Backward Europe and Advanced Asia, Lenin, Works, vol. 19).

deceitful, power-hungry, loving, suffering, insecure, courageous, cheerful, and much more. Some resented the mirror he held up to them; others, in his brilliant analyses of the history of science, recognized themselves and their peers, and in doing so also saw that scientific rules must be broken if we are to gain knowledge. The idea of a fixed method and the insistence on a rigid theory is not only naive, but ideological, and the disregard for other sources of knowledge is a sign of narrow-mindedness and chauvinism (Feyerabend, 1986, p. 392).

References

Baum, W. (Hrsg.). (1997). *Paul Feyerabend—Hans Albert. Briefwechsel.* Fischer.

Der Spiegel (vom 26.05.1968). Liebe anders. https://www.spiegel.de/politik/die-kommunisten-sind-uns-zu-buergerlich-a-0735005a-0002-0001-0000-000046039805. Accessed: March, 25, 2023.

Feyerabend, P. K. (1975). *Against method. Outline of an anarchistic theory of knowledge.* New Left Books.

Feyerabend, P. K. (1976). Logic, literacy, and Professor Gellner. *The British Journal for the Philosophy of Science, 27*(4), 381–391.

Feyerabend, P. K. (1977). Unterwegs zu einer dadaistischen Erkenntnistheorie. In H. P. Duerr (Hrsg.), *Unter dem Pflaster liegt der Strand* (S. 9–88). Karin Kramer Verlag.

Feyerabend, P. K. (1978a). *Science in a free society.* New Left Books.

Feyerabend, P. K. (1978b). *Der wissenschaftstheoretische Realismus und die Autorität der Wissenschaften.* Vieweg & Sohn.

Feyerabend, P. K. (1980). *Erkenntnis für freie Menschen.* Suhrkamp.

Feyerabend, P. K. (1986). *Wider den Methodenzwang.* Suhrkamp.

Feyerabend, P. K. (1995a). *Briefe an einen Freund, herausgegeben von Hans Peter Duerr.* Suhrkamp.

Feyerabend, P. K. (1995b). *Zeitverschwendung.* Suhrkamp.

Hörz, H. (1983). Natur, Naturwissenschaften, Kultur. *Deutsche Zeitschrift für Philosophie, 31*(7), 785–799.

Hörz, H. (2019). Naturerkenntnis und Gesellschaftsgestaltung. *Sitzungsberichte der Leibniz-Sozietät der Wissenschaften zu Berlin, 139–140,* 291–297.

Jung, J. (2000). Paul K. Feyerabend. Last Interview. In J. Preston, C. Munévar, & D. Lamb (Eds.), *The worst enemy of science. Essays in memory of Paul Feyerabend.* Oxford University Press.

Lenin, W. I. (1968; Original: 1913). *Werke,* Vol. 19. Dietz.

Martin, E. C. (2019). "The battle is on": Lakatos, Feyerabend, and the student protests. *European Journal for Philosophy of Science, 9*(2), 28.

Marx, K. (1983; Original: 1857). *Einleitung zu den „Grundrissen der Kritik der politischen Ökonomie." Karl Marx & Friedrich Engels, Werke,* Vol. 42. Dietz.

Oberheim, E. (1997). Bibliographie Paul Feyerabends. *Journal for General Philosophy of Science/Zeitschrift für allgemeine Wissenschaftstheorie, 28*(1), 211–234.
Trotzki, L. (2016; Original: 1921). Die Schule der revolutionären Strategie. In L. Trotzki, *Europa und Amerika*. Mehring Verlag.

Found Object: Philosophy of Nature 8

> *"The aim of primitive myths and scientific theories is exactly the same: to grasp and causally explain striking phenomena. The method is also the same—it consists in drawing conclusions from the available observational material" (Feyerabend, 2009, p. 91).*

Feyerabend wrote "Against Method" within a year and later remarked that with it he had said everything he ever wanted to say. However, the harsh criticisms that followed its publication took a toll on him.

"Some time after the storm of indignation had arisen, I fell into a depression that lasted for over a year. The dejection was like an animal, spatially bounded and locatable. I would wake up, open my eyes and listen: is it there, or is it not there? Not a trace of it. Maybe it's still asleep. Maybe it will leave me alone today. Cautiously, very cautiously, I slip out of bed. Everything is quiet. I go to the kitchen and start to have breakfast [...] Just a short trip to the bathroom, pull on the sweater, and out I go for the morning walk.—And there it is again, my clingy depression..." (1995, p. 199).

No wonder, the objections from his critics were at times massive. For example, David R. Topper (1975, p. 384) considers Feyerabend's book brilliant and insightful, but claims that with his conclusions the author sides with the anti-scientific gurus. Feyerabend's good acquaintance Ernest Nagel reproaches him that his discussions on the incommensurability of theories are not very helpful and that his conception of knowledge leads to absurd conclusions (Nagel, 1977, p. 1134). Klaus Fischer opposes Feyerabend's view that Galileo was a methodological anarchist. Rather, Galileo had clear ideas about proper scientific work and

W. Frindte, *Against Narrow-Mindedness and Chauvinism—Journeying Through Absurd Times with Paul K. Feyerabend*, https://doi.org/10.1007/978-3-658-50113-6_8

argumentation (Fischer, 1992, p. 192). Marxists accused Feyerabend of wanting to make the teaching of "magic" and "anti-revolutionary knowledge" accessible in schools (Ley, 1981, p. 379), and that his "anarchist epistemology" was merely an "extreme form of positivism" that had turned into an "apology for myth" and "philosophy of life" (Gedö, 1979, pp. 1468, 1472). Herrmann Ley (1911–1990) apparently misunderstood Feyerabend. In the passage in question to which Ley refers, Feyerabend *does not* demand "for democratic reasons the admission of the teaching of magic in schools" (Ley, 1981, p. 379), but rather emphasizes that it is important for a mature citizen (male or female) to study science, but also the fairy tales and myths of "primitive" societies in preparation for his (her) profession, "in order to acquire the knowledge necessary for a free decision" (Feyerabend, 1986, p. 396).

A comprehensive and very personal critique of "Against Method" came from the London School of Economics and Political Science (LSE): Ernest Gellner, professor of philosophy at the LSE, not only attacked Feyerabend's arguments, but above all took aim at the author himself. In contrast to Imre Lakatos, who always strove to meet the highest scientific standards both in his writings and in his lectures (Gellner, 1975, p. 331), Feyerabend was, according to Gellner, a clown whose clownery was persistently rough, boastful, mocking, arrogant, aggressive, and egotistical. His frivolity contained a distinctly sadistic streak, which manifested itself in the obvious pleasure he took in confusing and intimidating the "rationalists" (ibid., p. 342). Paul Feyerabend responded to "Dear Professor Gellner" a year later. The reviewer, meaning Gellner, writes well, but by no means correctly. He is strikingly blind to the ideas, motives, and approaches of others. Gellner suffers from reading and comprehension errors; his critique of "Against Method" is a failed attempt at rational criticism (Feyerabend, 1976, p. 382).

Helmut Spinner even drew a comparison with Adolf Hitler. Feyerabend's naive philosophical revolutionism, he claimed, finds its exact counterpart in Hitler's well-thought-out political philosophy, which teaches that nothing is impossible, that in politics only the principle of absence of principles applies. "Just as Feyerabend has handed science over to an unrestrained principle of opportunism, so Hitler handed over politics..." (Spinner, 1977, p. 573). In "Erkenntnis für freie Menschen" (1980, pp. 105; "Science in a Free Society", 1978) Feyerabend responds to Spinner. That Hitler was an opportunist, Feyerabend essentially argues, is hardly suitable for establishing such a connection between "Against Method" and National Socialism; Einstein and Bohr, as far as their scientific work was concerned, were also ruthless opportunists and admitted as much. The

Hitler problem, Feyerabend continues, probably weighs more heavily than little Helmut (meaning Spinner) can imagine (Feyerabend, 1980, pp. 106).

Joseph Agassi, his almost-friend, did find some interesting things in "Against Method," but also said it was hard to ignore the nonsense and focus on the valuable material in the book. Feyerabend, he said, commented on all sorts of social scientific, artistic, and historical events, but did not say a word about National Socialism, fascism, or racism. Despite the violence and vulgarity he found in the book, Agassi decided to write a review, also because he hoped that Feyerabend would rid himself of the rubbish and finally become the good-natured and exciting scientist he so wanted to be (Agassi, 1976, pp. 173).

It may be that Feyerabend, at least until the publication of "Against Method," had little interest in an engaged confrontation with *political* world events and their history.

But, as mentioned, the critical reactions to his book did not leave Feyerabend unaffected. Perhaps he already sensed, while working on "Against Method," what might be in store for him. And possibly this is also a reason why, in parallel to this book, he worked on a manuscript with which he attempted to "correct the […] self-staged image of a frivolous thinker" (Heit & Oberheim, 2009, p. 13). In 2004, Helmut Heit and Eric Oberheim discovered a book manuscript on natural philosophy in the Feyerabend estate in Konstanz. Feyerabend had begun working on this manuscript in the early 1970s and later apparently revised it with the intention of publishing it in 1976. A total of three volumes were planned. The fact that this did not come to pass, Heit and Oberheim suspect (2009, p. 14), also had to do with the revisions Feyerabend wanted to make to his "Against Method." The manuscript on natural philosophy was not published during his lifetime and was only thankfully edited by Heit and Oberheim in 2009.[1] At its core, Feyerabend's aim in this book is to demonstrate how the rise of rationalism unfolded from early times through antiquity to the modern era. In doing so, Feyerabend draws, among other things, on research from prehistoric studies, on ancient Egyptian and Babylonian art and science, on interpreters of the Homeric epics, on social-anthropological studies of indigenous peoples, and on works concerning "Western" natural philosophy from Parmenides, Xenophanes, Aristotle through Bacon, Galileo, Newton, Descartes, Leibniz, Hegel, and Mach to Niels Bohr. Konrad Lorenz, Jean Piaget, and Noam Chomsky are invoked to point to more or less innate internal schemata that help us find our way in the world (Feyerabend, 2009, p. 154).

[1] Reviews of this volume can be found, among others, in Oren Harman (2019), Ian James Kidd (2019), and Thomas Kupka (2011).

In doing so, he succeeds in showing that "the testimonies of early and ancient cultures are just as theory-dependent and at the same time partially successful conceptions of reality as our own…" (Heit & Oberheim, 2009, p. 29). This also applies to the myths of earlier cultures. These myths are not mere fabrications; rather, they serve to capture and causally explain striking events (Feyerabend, 2009, p. 91). Scientific theories and myths are the results of human activities and can serve as guidelines for human actions. Their functionality or practicality is not shown in their truth value, but in whether and to what extent they can contribute to promoting social progress within a community. Moreover, it must be admitted that "scientific theories, myths, and other results of human activity, as results of *human* activity, reflect human characteristics. […] It must further be admitted […] that the 'facts' of the world are not merely *mirrored* by human ideology, but are at least partially *constructed* by it" (Feyerabend, ibid., p. 102; emphasis in the original). Does this already indicate a certain degree of constructivism in Feyerabend's views (see also *Chap. 12*)?

The manuscript "Natural Philosophy" concludes with an optimistic outlook, which Feyerabend intended to elaborate in the subsequent volumes—volumes that, unfortunately, never materialized: On the horizon, he writes, a new philosophical-mythological science is emerging—still indistinct—in which sympathy with nature, an intuitive understanding of the manifold forms of life, and the full development of one's own personality will be essential components (Feyerabend, 2009, pp. 324 f.).

Be that as it may: Heit and Oberheim point out that Feyerabend spent many years explaining his arguments in "Against Method." "Perhaps *Philosophy of Nature* would have been a better answer to critics than *Science in a Free Society* was" (Heit & Oberheim, 2009, p. 15; emphasis in the original).

References

Agassi, J. (1976). Against method: Outline of an anarchistic theory of knowledge. *Philosophia, 6*(1), 165–177.

Feyerabend, P. (1976). Logic, literacy, and Professor Gellner. *The British Journal for the Philosophy of Science, 27*(4), 381–391.

Feyerabend, P. K. (1980). *Erkenntnis für freie Menschen*. Suhrkamp.

Feyerabend, P. K. (1986). *Wider den Methodenzwang*. Suhrkamp.

Feyerabend, P. K. (1995). *Zeitverschwendung*. Suhrkamp.

Feyerabend, P. K. (2009). *Naturphilosophie*, herausgegeben von H. Heit und E. Oberheim. Suhrkamp.

Fischer, K. (1992). Die Wissenschaftstheorie Galileis—oder: Contra Feyerabend. *Journal for General Philosophy of Science / Zeitschrift für allgemeine Wissenschaftstheorie, 23*(1), 165–197.

Gedö, A. (1979). Positivismus und „Postpositivismus". *Deutsche Zeitschrift für Philosophie, 27*(12), 1467–1474.

Gellner, E. (1975). Beyond truth and falsehood. *The British Journal for the Philosophy of Science, 26*(4), 331–342.

Harman, O. (2019). Philosophy of nature by Paul Feyerabend. *Common Knowledge, 25*(1), 458–459.

Hattiangadi, J. N. (1977). The crisis in methodology: Feyerabend. *Philosophy of the Social Sciences, 7*(3), 289–302.

Heit, H., & Oberheim, E. (2009). Paul Feyerabend als historischer Naturphilosoph. Einführung. In P. Feyerabend (Hrsg.), *Naturphilosophie* (pp. 7–37). Suhrkamp.

Kidd, I. J. (2019). Philosophy of nature, written by Paul Feyerabend. *Journal of the Philosophy of History, 13*(2), 281–285.

Kupka, T. (2011). Feyerabend und Kant: Kann das gut gehen? Paul K. Feyerabends Naturphilosophie und Kants Polemik gegen den Dogmatismus. *Journal for General Philosophy of Science, 42*(2), 399–409.

Ley, H. (1981). Erbe, Rezeption und ideologischer Klassenkampf. *Deutsche Zeitschrift für Philosophie, 29*(3), 367–379.

Nagel, E. (1977). Reviewed work(s): Against method: Outline of an anarchistic theory of knowledge by Paul Feyerabend. *The American Political Science Review, 71*(3), 1132–1134.

Spinner, H.F. (1977). Thesen zum Thema Reichweite und Relevanz der Wissenschaftstheorie für die Einzelwissenschaften—Analytische Philosophie versus Marxismus. In K.-H. Braun & K. Holzkamp (Eds.), *Kritische Psychologie, Bericht über den I. Internationalen Kongress Kritische Psychologie,* Bd. 2. Pahl-Rugenstein.

Topper, D. R. (1975). Feyerabend, „Against Method" (Book Review). *Canadian Journal of History/Annales Canadiennes d'Histoire, 10*(3), 393–395.

Knowledge for Free People

9

> *"The answer that I give, explain, and defend in the present book is as follows: in a free society, a citizen uses the standards of the tradition to which he belongs; he uses Hopi standards if he is a Hopi Indian; fundamentalist Protestant standards if he belongs to the sect of the fundamentalists; ancient Jewish standards if he wishes to revive ancient Jewish traditions; fascist standards if fascism is closer to him; then there are special groups with special interests and ideas, such as the women's movement, the gay movement, ecological groups, and many more" (Feyerabend, 1980a, pp. 12; emphasis in the original).*

Three years after "Against Method," Paul Feyerabend followed up with "Science in a Free Society" (Feyerabend, 1978). Now he was calling not only for a *scientific* pluralism, but for the "equal rights of all traditions" in a free society (Feyerabend, 1980a, p. 18).[1]

As is well known, society was by no means as free and peaceful as Paul Feyerabend wished it to be at the end of the 1970s. Not in the USA, not in West Germany, nor in Eastern Europe, Africa, South America, or Asia.

---

[1] This German edition, "Erkenntnis für freie Menschen" ("Knowledge for Free People"), is a revised version of the English original. Compared to the original, Feyerabend made additions, changes, and also cuts, which will not be discussed further here. However, a note on "pluralism" is still in order. Paul Feyerabend resisted being called the "inventor of theoretical pluralism." "In modern times, it was above all Mill who showed how knowledge can be advanced through a competition of alternatives. His conception of method is clearer than Popper's and is closer to scientific practice" (Feyerabend, 1980b, pp. 191).

W. Frindte, *Against Narrow-Mindedness and Chauvinism—Journeying Through Absurd Times with Paul K. Feyerabend*,
https://doi.org/10.1007/978-3-658-50113-6_9

Overview

In 1977, the civil war began in Mozambique and the Ogaden War between Ethiopia and Somalia. The USA and the Soviet Union used these and other conflicts in Africa as a "playground" for the global Cold War. Also in 1977, RAF terrorists ("Red Army Faction") murdered the Federal Prosecutor General Siegfried Buback, the spokesman of the board of Deutsche Bank Jürgen Ponto, and the president of the employers' association Hanns Martin Schleyer. In October, the Lufthansa plane *Landshut* was hijacked by a Palestinian terrorist commando. After the passengers taken hostage were freed by soldiers of the Federal Border Guard, the RAF terrorists of the so-called first generation, who were imprisoned in Stammheim, took their own lives. In March 1978, over 20,000 Israeli soldiers occupied southern Lebanon to put an end to the terrorist acts of the PLO.

In September 1978, Israeli Prime Minister Menachem Begin and Egyptian President Anwar al-Sadat agreed at Camp David to initiate peacemaking measures in the Middle East. In the spring of 1979, Islamic revolutionaries overthrew the hated monarchy of Shah Pahlavi in Iran and founded the Islamic Republic, based on violence. In June 1979, the USA and the Soviet Union did sign the SALT II treaty on arms limitation. At the same time, the Soviet Union was expanding its nuclear arsenal, including in the GDR. And the US Department of Defense developed a new strategy for nuclear warfare, which culminated at the end of 1979 in the NATO Double-Track Decision and led to the deployment of new nuclear intermediate-range missiles in Western Europe. In July 1979, troops of the Sandinista National Liberation Front ended the dictatorial regime of Somoza in Nicaragua. Two years later, supporters of Somoza, with the support of the US government under Ronald Reagan, began a civil war against the Sandinistas. In November 1979, Iranian students occupied the US embassy in Tehran and took the diplomats hostage; in December, the Soviet invasion of Afghanistan began, and so on and so forth.

Paul Feyerabend is an intelligent and alert contemporary. It can be assumed that he was aware of the social, political, military, and cultural developments in these turbulent times. In an analysis of Feyerabend's views on the relationship between science, ideology, and the Cold War, Ian James Kidd (2016) reaches similar conclusions. Feyerabend's call for a free society must also be seen against the background of the ideological confrontations between the USA and the USSR. On the one hand, these confrontations may have encouraged Feyerabend to advocate even more strongly than before for a socially engaged science. On the other hand, he was suspicious of a "strict dichotomy" in which one had to be either on the side of the USA and its democracy or on the side of the USSR and communism. Such a dichotomy would be at odds with his radically critical stance toward any ideology and toward an ideological monism that is based on a *single* authoritative vision of society, history, and reality (Kidd, 2016, pp. 71).

To what extent these social backgrounds also influenced the work on "Science in a Free Society" cannot be said with certainty, but it is likely. His plea for a *democratic relativism* could be evidence of this. According to Feyerabend, at least three arguments speak in favor of such relativism:

First: "People have the right to live as they see fit, even if their lives appear stupid, bestial, obscene, or godless to others" (Feyerabend, 1980a, p. 17). No institution, for example, should be allowed to determine whether and how people wish to receive medical treatment, even if the forms of treatment contradict Western medicine.

Second: The basic idea here is "that a society containing many traditions provides its citizens with better means of judging these traditions than a society with a single basic ideology" (ibid., p. 19).

Third: Scientific standpoints, ideas, and procedures are not only incomplete but also error-prone. Since fundamental disputes about traditions are disputes between laypeople, the outcome of these disputes *"cannot be subjected to any higher judgment* [than] *again the judgment of laypeople"* (Feyerabend, 1980a, pp. 20; emphasis in original).

Democratic relativism not only supports a right, it is also a "highly useful research instrument for any state that makes it its basic philosophy" (ibid., p. 22). And, to anticipate the quintessence of his arguments in "Erkenntnis für freie Menschen" ("Science in a Free Society"), Feyerabend formulates the slogan *"citizens' initiatives instead of philosophy"* (ibid., p. 23; emphasis in original) and, a few pages later, *"citizens' initiatives instead of epistemology"* (ibid., p. 37). The structural core—if one can say that against the background of Feyerabend's anarchist epistemology—are traditions and their functions. However, it would contradict the epistemological anarchism of Feyerabend's persuasion to expect from him an exact definition of tradition that conforms to scientific rules. He does not provide such a definition; the terms Feyerabend juggles with are usually those with a broad and often diffuse range of meanings. The brief note hidden in a footnote of a 1977 work does not help much further. There, Feyerabend writes: "'Tradition' is here used in a wide sense, covering social, psychological as well as bio-physiological phenomena" (Feyerabend, 1977, p. 11).

In order to criticize commonly used concepts and examine their usefulness, a standard of assessment is necessary, which—as one reads in "Against Method"—can be derived either from a new theoretical system or from another science, from religion, from mythology, or from the outpourings of madmen (Feyerabend, 1986, p. 87). This also means that the concepts are not judged according to truth, falsehood, or logical consistency, but according to their usefulness in the context of other concepts of the new theoretical or pre-theoretical system. Concepts, Feyerabend writes in a later work, "are ambiguous, elastic, capable of reinterpretation, extrapolation, restriction" (Feyerabend, 1987, p. 78). Thus, Feyerabend uses the concept of *tradition* among others as synonymous with *practice, reason,* or *science* (e.g., Feyerabend, ibid., pp. 62, 66, 195).[2] Traditions, it can be said, are for him not only symbolic, written, and instrumental memories of a social community, but, as Arno Waschkuhn puts it (1999, p. 245), "human-social constructs of a sociocultural kind" that are available for practical application and social identification. Traditions can be of an abstract nature, such as astronomy, mathematics, or the laws of physics. "Abstract traditions were first introduced in the West by the pre-Socratics" (Feyerabend, 1980a, p. 65), in order to derive new laws and rules and to anticipate future procedures.

Traditions with local laws, Feyerabend calls *historical* traditions.

"Traditions are neither good nor bad; they simply exist. 'Objectively,' that is, independently of traditions, there is no choice between a humanitarian attitude and anti-Semitism [...]. *A tradition acquires desirable and undesirable features only when it is related to a tradition, that is, when it is considered as a participant in a tradition and judged according to the values of that tradition* [...]. There are therefore at least two different ways in which collective problems can be brought to a solution. I call these ways a *free exchange* (of thoughts, goods, actions, etc.) and a *guided exchange* [...]. The participants in a guided exchange accept a tradition and allow only those actions (considerations, arguments, procedures) that conform to the standards of this tradition [...]. A free exchange respects all features of the opponent, whether an individual or a nation [...]. *A free society is a society in which all traditions have equal rights and equal access to the centers of education and other centers of power"* (Feyerabend, 1980a, pp. 68; emphasis in the original).

[2] In his autobiography, Feyerabend confesses, among other things, that he wanted to free people from the tyranny of "abstract concepts" such as "truth," "reality," or "objectivity," but that he himself introduced "similarly rigid concepts" like "democracy" or "tradition," and wonders how this could have happened (Feyerabend, 1995, p. 246).

Paul Feyerabend is clever enough to anticipate the objections that can be raised against his theses—ethical, political, and scientific objections. He dismisses all three. Ethical objections, he says, are often presented with examples that attempt to put potential supporters of his relativism under moral pressure: Hitler, the Second World War, Auschwitz, and more recently terrorism, are brought up with tiresome regularity. Political objections concern, among other things, the question, "what the author intends to do should conservative or fascist tendencies gain the upper hand" (ibid., p. 76). The scientific objections are the most foolish, because they are based on the assumption that scientific methods are superior to all other methods and rules.

But how can a "red line" be drawn so as not to grant all traditions equal rights and equal access to the centers of power?

It is not necessary to refer to the Shoah to show that the (traditional) social constructions of a community (e.g., German antisemites) *must not* have the same rights as those of other communities (e.g., Jews). Perhaps it is enough to find a *cross-cultural* tradition or social construction that could make it possible to choose between a humanitarian attitude and anti-Semitism. Such a *cross-cultural* tradition could be an instruction known as the *Golden Rule,* which can be expressed in the proverb: "Do not do to others what you do not want done to yourself."

This is a principle that probably only received its name as the *Golden Rule* in the 17th and 18th centuries (Hruschka, 2004, p. 162). In the era of the European Enlightenment, it is discussed, among others, by the Englishman Thomas Hobbes (1588–1679) or by the German philosopher Christian Thomasius (1655–1728), criticized by Immanuel Kant (1724–1804), and reformulated as a rule of prudence in his "categorical imperative." The "categorical imperative" famously states: "Act only according to that maxim whereby you can at the same time will that it should become a universal law" (Kant, AA, vol. IV, p. 421; original: 1785). Paul Feyerabend believes that Kant constructed a "cruel dream world" with the imperative (Feyerabend, 1992, p. 74; original: 1989).

As a principle, however, the *Golden Rule* has a longer history. According to experts, it is found, albeit in different formulations, in Chinese, Indian, Greek, Roman, Islamic, and Jewish writings—mostly as an everyday and/or philosophical or religious moral law, among others in the sayings of Thales, in the historical work of Herodotus, in the Babylonian Talmud (formulated by the great teacher Hillel the Elder, probably died 10 CE),[3] by the Apostle Matthew[4] or by the philosopher Augustine of Hippo (Reiner, 1977, pp. 232).

Paul Feyerabend is, as is well known, familiar with the Greek thinkers. Thales or Herodotus are not foreign to him. Although he rejects the insistence on moral principles as moral pressure, he considers it possible to use historical sources not only to formulate scientific problems, but, if necessary, also to solve them. Perhaps, one might now object, the historical sources to which the *Golden Rule* refers could also serve to decide between "good" and "bad" traditions after all. Such an objection would not contradict Feyerabend's argument that traditions ultimately decide everything, or at least much. The Golden Rule is not a scientific method, but a traditional principle with which, so to speak, a "red line" could be drawn in order not to grant all traditions equal rights and equal access to the centers of power. Certainly, every rule has its limits, including the Golden Rule. How can these limits be recognized and secured?

Feyerabend's answer: *"Not rationalist standards, not religious convictions, not humane impulses, but citizens' initiatives are the filter that separates useful from useless ideas and measures"* (Feyerabend, 1980a, p. 77; emphasis in the original). The citizens of a region, a city, or a village—and not clueless intellectuals—should decide in a free society about the value and use of ideas (ibid., p. 104).

But what happens if citizens themselves are so entrenched in their traditions that they are virtually unable to move beyond the boundaries of their own self-created and inherited sociocultural constructions, and seek the frames of reference for their initiatives within what we would now call their group-specific echo chambers? Feyerabend also envisions such transgressions of boundaries (Feyerabend, 1980a, p. 93). Rational research is often only temporarily useful (ibid., p. 99), and the "tradition(s) of the white man" (p. 127) cannot serve as a universal guideline. These traditions have brought too much harm into the world.

---

[3] Lazarus Goldschmidt translated the words of Hillel from the Babylonian Talmud as follows: "What is hateful to you, do not do to your neighbor. That is the whole Torah and the rest is only commentary; go and learn it" (Goldschmidt, 1929–1936, p. 522, Sabbath II, v).

[4] "So in everything, do to others what you would have them do to you; for this sums up the Law and the Prophets" (Matthew 7:12).

The "[...] democratic principles as proclaimed by white men (and women), as they are practiced today, are incompatible with the undisturbed existence, development, and growth of specific cultures" (ibid., p. 129). It is no less narrow-minded to reject "non-scientific" developments and traditions, such as traditional medical diagnostic and healing methods (Feyerabend, 1980a, p. 204, mentions acupuncture among others), simply because, although they may help sick people, they can hardly be explained by Western academic medicine.

The more than 3,000-year-old Chinese treatment method is still controversially discussed by both laypeople and experts. In Germany, over 15,000 physicians are reportedly registered with the additional qualification "acupuncture" at the relevant chambers. Whether acupuncture, for example, can help with chronic pain or whether the effects reported by patients are merely a placebo effect still seems to be a question that cannot be fully answered. Meta-analyses—that is, studies that quantitatively synthesize earlier empirical individual analyses—do show, however, that acupuncture can, for example, contribute to pain relief in rheumatoid arthritis or serve as an effective supplement to conservative (Western) therapies (e.g., Huo, et al., 2021; Li, et al., 2022). It is striking, however, that the vast majority of such meta-analyses have been conducted by Chinese researchers.

But where are the limits of a democratic (and scientific) relativism? Feyerabend not only distinguishes between abstract and historically local traditions and calls for the "equal rights of all traditions" (see above). He also sees a difference between "opportunistic or eclectic traditions" on the one hand and "dogmatic traditions" on the other. While the eclectic traditions are guided by values but do not hesitate to change their own values and to tolerate the values of other traditions, the dogmatic traditions or their representatives attempt to project their own values as the only true fundamental values onto the world and "[...] to measure all events (of history, of private life, even of nature) by them and to try to bend the world in their direction through violence, persuasion, or institutional machinations..." (1980a, p. 136).

Feyerabend's sympathies, as can be inferred from his arguments, undoubtedly lie with the eclectic traditions. They correspond to his relativism and enable the "[...] *opportunistic adoption* and alteration of what is useful (whereby the criteria of usefulness change from problem to problem and from era to era)" (ibid., p. 140; emphasis in the original). Thus, it is not an absolute relativism that

Feyerabend demands, but rather the support of and participation in eclectic traditions. Dogmatism, in whatever guise, is to be rejected. Dogmatism lies beyond the boundary that democratic relativism does not cross.[5] However, whether a "police force *from outside,* which restricts *physical* freedom of movement but not the *flight of thought*" (1980a, p. 296; emphasis in the original), is capable of restricting the freedom of movement of dogmatic movements may rightly be doubted and can probably only be interpreted as the excuse of an author who was still somewhat politically naïve into the early 1980s.[6]

However—and here one must again agree with Paul Feyerabend—it is "[…] short-sighted to assume that one has 'solutions' for people whose lives one does not participate in and whose problems one does not know" (1980a, p. 237).

"What I write, what I tell other people, what I bore my friends with, is not a 'conception of knowledge,' but a collection of hints, suggestions, aphorisms that illuminate certain situations and are meant to help the reader or listener think through their problems. The content of the remarks is always the same: do not trust the sciences, do not trust the intellectuals, whether they are Marxists or right-wing Catholics, they have their own interests, they want to gain spiritual and material power over all people […]. Whom do I have in mind when I make such remarks? I do not think of intellectuals, I think of people who have, or once had, their own tradition, to whom this tradition once gave meaning and identity, but whose traditions have been ridiculed and dismissed as laughable prejudices; they are denied reason, knowledge, insight, they are denied everything that is fashionable today. I think of the Hopi in America, of Black people who are slowly succeeding in rediscovering and reviving their own cultures, which had almost completely disappeared. These are the people I think of, I think that they have

[5] "Psychologically speaking, dogmatism arises, among other things, from the inability to imagine alternatives to the point of view in which one believes," writes Feyerabend in "Problems of Empiricism" (Feyerabend, 1962, p. 69).

[6] That Paul Feyerabend, over the course of the 1980s and early 1990s, looked at sociopolitical developments with increasing interest, knowledge, and critical awareness was certainly also due to the influence of Grazia Borrini-Feyerabend. Some politically and historically significant books that appeared during this period, which Feyerabend read very thoroughly and annotated extensively, likely also illustrate his heightened sociopolitical interest (see: Paul Feyerabend Archiv, University of Konstanz). For example, "Death is a Master from Germany" by Lea Rosch and Eberhard Jäckel (1990), "Dialectics without Dogma?" by Robert Havemann (1990; original: 1964), or (which greatly surprised me) the book "Carl Zeiss—The Adventurous History of a German Company" by Armin Hermann (1992).

a right to live as they please, and I also think that, fundamentally, there is no 'rational' argument to prevent them from living this way" (Feyerabend, 1980a, p. 284).

One could add that the freedom of a society is measured not only by the freedom of movement of its various traditions, but also by the individual and social possibilities of drawing on different traditions, not just one's own or those of one's own group. Thus, in a free society, for example, it should be possible for a Hopi person to demand not only Hopi standards, but also Protestant, ancient Jewish, and other traditions for shaping their own and social life. The social existence, freedom of movement, and access to such and other traditions are undoubtedly among the *possibilities* that people can orient themselves by when they try to construct their social reality. But rituals, conventions, myths, social norms, the diversity of social relations, the economic, political, cultural, and scientific structures and processes also offer opportunities for individual development, individual worldviews, for individual constructions of social realities, and for social reality constructions.

This does not say everything about knowledge for free people, but perhaps it says the most important thing.

References

Feyerabend, P. K. (1977). Rationalism, relativism and scientific method. *Philosophy in Context, 6*(1), 7–19.

Feyerabend, P. K. (1978). *Science in a free society.* New Left Books.

Feyerabend, P. K. (1980a). *Erkenntnis für freie Menschen.* Suhrkamp.

Feyerabend, P. K. (1980b). Eine Lanze für Aristoteles. In G. Radnitzky & G. Andersson (Eds.), *Fortschritt und Rationalität der Wissenschaft.* (pp. 157). J. C. B. Mohr.

Feyerabend, P. K. (1986). *Wider den Methodenzwang.* Suhrkamp.

Feyerabend, P. (1987). Putnam on incommensurability. *The British Journal for the Philosophy of Science, 38*(1), 75–81.

Feyerabend, P. K. (1962). Explanation, Reduction, and Empiricism. In H. Feigl & G. Maxwell (Eds.), *Minnesota Studies in the Philosophy of Science,* Vol. 3. University of Minnesota Press.

Feyerabend, P. K. (1992; Original: 1989). *Über Erkenntnis. Zwei Dialoge.* Campus.

Feyerabend, P. K. (1995). *Zeitverschwendung.* Suhrkamp.

Goldschmidt, L. (1929–1936). Der Babylonische Talmud, Band II. Jüdischer Verlag. https://archive.org/details/DerBabylonischeTalmudLazarusGoldschmidt19291936/mode/2up. Accessed: Febr. 15, 2023.

Havemann, R. (1990; Original: 1964). *Dialektik ohne Dogma?* Akademie-Verlag.

Hermann, A. (1992). *Carl Zeiss – die abenteuerliche Geschichte einer deutschen Firma.* Piper.

Hruschka, J. (2004). Die Goldene Regel in der Aufklarung die Geschichte einer Idee. *Jahrbuch für Recht und Ethik, 12,* 157–172.

Huo, X., Liang, L., Ding, X., Bihazi, A., & Xu, H. (2021). Efficacy and safety of acupuncture with western medicine for rheumatoid arthritis: A systematic review and meta-analysis. *Acupuncture & Electro-Therapeutics Research, 46*(4), 371–382.

Kant, I. (1785). Grundlegung zur Metaphysik der Sitten. In I. Kant (Eds.), *Gesammelte Schriften* (Vol. IV, pp. 385–463, Vol. 1–22 Preußische Akademie der Wissenschaften, Vol. 23 Deutsche Akademie der Wissenschaften zu Berlin, ab Vol. 24 Akademie der Wissenschaften zu Göttingen). Walter de Gruyter & Co.

Kidd, I. J. (2016). "What's so great about science?" – Feyerabend on science, ideology, and the cold war. In E. Aronova & S. Turchetti (Eds.), *Science studies during the cold war and beyond* (pp. 55–76). Palgrave Macmillan.

Li, H., Man, S., Zhang, L., Hu, L., & Song, H. (2022). Clinical efficacy of acupuncture for the treatment of rheumatoid arthritis: Meta-analysis of randomized clinical trials. *Evidence-Based Complementary and Alternative Medicine;* https://www.hindawi.com/journals/ecam/2022/5264977/. Accessed: Febr. 15, 2023.

Paul Feyerabend Archiv. Universität Konstanz. https://www.kim.uni-konstanz.de/phil-archiv/bestaende/paul-feyerabend/. Accessed: Febr. 7, 2023.

Reiner, H. (1977). Die Goldene Regel und das Naturrecht: Zugleich Antwort auf die Frage: Gibt es ein Naturrecht? *Studia Leibnitiana, 9*(2), 231–254.

Rosch, L. & Jäckel, E. (1990). *Der Tod ist ein Meister aus Deutschland.* Hoffmann und Campe.

Waschkuhn, A. (1999). Die Loslösung bei Paul Feyerabend: Erkenntnis für freie Menschen. In derselbe, *Kritischer Rationalismus* (pp. 233–254). Oldenbourg Verlag.

10 The Destruction of Diversity

"How can it happen that perspectives which diminish richness and deprive human existence of its value become so powerful?" (Feyerabend, 2005, p. 39).

Paul Feyerabend wrestled with this question until the end of his life. In 1999, Feyerabend's work "Conquest of Abundance. A Tale of Abstraction versus the Richness of Being" was published posthumously, edited by Bert Terpstra (Feyerabend, 1999).[1] The German translation bears the title "Die Vernichtung der Vielfalt. Ein Bericht" (Feyerabend, 2005; "The Annihilation of Diversity. A Report"). The book is divided into two parts. The first part contains a fusion of three versions of a manuscript on which Feyerabend worked until his death. The second part features a selection of essays that had already appeared elsewhere.

In the foreword to the German edition, Grazia Borrini-Feyerabend writes that in the book, Paul Feyerabend recounts some special moments in the development of Western culture—periods in which complex worldviews with their overflowing interpretations of reality "had to give way to a few abstract concepts and stereotypical representations" (Borrini-Feyerabend in: Feyerabend, 2005, p. 14).

One of the main questions Feyerabend tries to answer in the first part of the book is precisely the one quoted at the beginning of this chapter. Feyerabend is not concerned with *that* the diversity of worldviews was destroyed over the course of history, but rather *how* it came about "that the rich, colorful, and overflowing world, which permeates us in so many ways, was divided into two

[1] This posthumously published book was critically reviewed by Helmut Heit (2006), Gonzalo Munévar (2002), Stephan M. Downes (2002), and several other authors.

W. Frindte, *Against Narrow-Mindedness and Chauvinism—Journeying Through Absurd Times with Paul K. Feyerabend*,
https://doi.org/10.1007/978-3-658-50113-6_10

realms, one of which still contains some life, while the other lacks almost all the properties and events that make our existence meaningful" (Feyerabend, 2005, p. 39).

From a psychological perspective, the question is why people do not simply accept the rich, colorful world as it is, but instead try to organize and simplify it. It is well known that people categorize their world (persons, objects, events, processes) in order to reduce the complexity of reality. Categorization means dividing items, objects, people into distinct (abstract) classes based on shared characteristic features. Such a process ultimately serves to reduce complexity in dealing with the rich and unmanageable facets of reality. Categorization means transforming the world into a more predictable and controllable place. Findings and considerations from evolutionary science suggest that categorizing the world guarantees a survival advantage for humans (generally for primates, but not for other mammals) (e.g., Cosmides & Tooby, 1992; Tomasello, 2019).

In the introduction to "Die Vernichtung der Vielfalt," Feyerabend (2005, pp. 27) mentions a psychological case described by the Soviet neuropsychologist Alexander R. Luria[2] in his book "The Mind of a Mnemonist" (Luria, 1968). The case concerns the journalist and memory artist Solomon Shereshevski[3], who, due to a synesthetic disposition, was able to store and reproduce many individual impressions, but was hardly able to select and categorize information. Shereshevski suffered from being unable to forget the manifold, overlapping sensory impressions. Therefore, he tried to convince himself that he simply no longer wanted to perceive the many sensory impressions. Feyerabend uses this example to point out that there seem to be situations and occasions in which people simply "block out" the diversity of their impressions and simplify reality.

However, Feyerabend is less interested in the *psychological* problem of abstraction and sensory diversity. He is more concerned with investigating how, over the course of history, structures (theology, philosophy, sciences) developed in which the "richness of the world" was denied and replaced by abstractions. "Distancing oneself from common sense and everyday experience, moving toward a world of abstract concepts, has brought advantages. But these advantages were distorted and, due to the fundamental inconsistencies of the whole enterprise, turned into a threat" (Feyerabend, 2005, p. 39).

---

[2] Not to be confused with the microbiologist Salvador Edward Luria, to whom Feyerabend also occasionally refers.

[3] Not "Sherashevsky," as is mistakenly written in "Die Vernichtung der Vielfalt."

Feyerabend seeks to understand this "enterprise," the "rise of rationalism," its advantages, transformations, and distortions. It is obvious that he focuses primarily on the emergence of philosophy and science in ancient Greece, for it was there—according to Western understanding—that a revolution in knowledge took place between 800 and 30 BCE. Thus, Feyerabend analyzes the language in the Homeric epics, the rationalism of the "conceited loudmouth" Xenophanes, the logical justification for the preservation of being in Parmenides and his influence on modern physics. Feyerabend takes the (supposed) discovery of central perspective by Filippo Brunelleschi (1377–1446) as an example to speculate on the relationship between relativism and realism. He embeds his case studies in the respective cultural, political, and technological contexts of the time, thus pointing to the ambiguity of supposed progress in philosophy, science, and art. People do not simply represent the world using abstract concepts; they also change the world and their categorical interpretations of it. "Since we are in the world, we do not merely imitate or assemble events, but we also recreate them whenever we imitate, thereby changing what we consider to be the enduring objects of our attention" (Feyerabend, 2005. p. 141).

There is no need to go into detail here about the selection of essays in the second part of the book "Die Vernichtung der Vielfalt." Only the final contribution, which had already appeared in 1994 in the journal "Common Knowledge," should be briefly mentioned. It is a short statement on a document signed, among others, by Gadamer, Derrida, and Rorty, which addresses "all parliaments and governments of the world" with the request to establish the study of philosophy and its history in all schools. Feyerabend comments on the letter, among other things, with the words: "The real problems of our time are not even touched upon. What are the real problems? They consist of war, violence, hunger, disease, and environmental catastrophes [...]. The philosophers and scientists who signed it (the appeal, WF) would have done better to issue a harsh condemnation of the crimes and murders happening in our midst, together with a call to all governments to intervene and stop the killing, if necessary by military force. Such a condemnation and such a call would have been understood. It would have shown that philosophy cares about its fellow human beings" (Feyerabend, 2005, p. 295).

In this respect, Helmut Heit's conclusion is to be agreed with, that in his last book Feyerabend appears as a thinker "who can imagine our world as a better, happier place, without being either pathetic or dogged in his plea for more diversity" (Heit, 2006, p. 619).

References

Cosmides, L., & Tooby, J. (1992). Cognitive adaptations for social exchange. In J. Barkow, L. Cosmides, & J. Tooby (Eds.), *The adapted mind: Evolutionary psychology and the generation of culture* (pp. 163–228). Oxford University Press.

Downes, S. M. (2002). Book Review: Conquest of abundance: A tale of abstraction versus the richness of being, the worst enemy of science? Essays in memory of Paul Feyerabend. *Science, Technology, & Human Values, 27*(1), 160–164.

Feyerabend, P. K. (1999). *Conquest of Abundance. A Tale of Abstraction versus the Tichness of Being, edited by Bert Terpstra.* The University of Chicago Press.

Feyerabend, P. K. (2005). *Die Vernichtung der Vielfalt. Ein Bericht, herausgegeben von Peter Engelmann.* Passagen Verlag.

Heit, H. (2006). Reviewed Work(s): Die Vernichtung der Vielfalt. Ein Bericht by Paul K. Feyerabend. *Zeitschrift für philosophische Forschung, 60*(4), 615–619.

Lurija, A. R. (1968; Original: 1965). *The Mind of a Mnemonist.* Basic Books.

Munévar, G. (2002). Critical Notice: Conquering Feyerabend's Conquest of Abundance. *Philosophy of Science, 69*(3), 519–535.

Tomasello, M. (2019). Becoming Human: A Theory of Ontogeny. Harvard Press.

Part IV
What Remains?

"Now let us assume that what I have written corresponds to the truth—does preaching this truth help to solve the problems of our time? Can it curb the mass murders that are taking place today in many countries? Can it eliminate intolerance, the lack of compassion and understanding, the narrow-minded selfishness of individuals, corporations, and institutions that have ruined our planet, who are aware of their crimes but have no intention of changing?" (Feyerabend, 1992, pp. 208).

References

Feyerabend, P.K. (1992; Original: 1989). Über Erkenntnis. Zwei Dialoge. Campus

11 "Those are indeed important questions"

"Whether our actions are 'relevant' is something we only ever know in retrospect, and even then it often takes a long time for the effects to become apparent" (Feyerabend, 1992, p. 209).

Three years after the first English edition of "Science in a Free Society" appeared, Feyerabend published two substantial volumes with the subtitle "Philosophical Papers" (Feyerabend, 1981a, b), in which he once again historically substantiates and explicates his views on the influence of theories on observations, on the incommensurability of theories, on critical rationalism, and on scientific (societal) pluralism and the role of traditions. Certainly, both volumes contain ideas that Feyerabend had already developed elsewhere (for example, already in his dissertation from 1951; see, for example, Feyerabend, 1981a, pp. 17). In his arguments, Feyerabend addresses the members of the scientific communities to which he himself once felt he belonged. However, as in "Science in a Free Society," it also becomes clear that he is concerned with more than this: with communication with people and communities outside established academic circles, who, due to their traditions, possess their own notions of reason and truth. These ideas are also summarized in "Farewell to Reason" (Feyerabend, 1987), likewise a collection of previously published works. Here, Feyerabend not only targets the distinction between the context of discovery and the context of justification, among other things; he also explicitly advocates for cultural diversity, which for him is the best defense against totalitarian rule.

The distinction between the context of discovery and the context of justification goes back to Hans Reichenbach, who belonged to the illustrious Vienna

W. Frindte, *Against Narrow-Mindedness and Chauvinism—Journeying Through Absurd Times with Paul K. Feyerabend*,
https://doi.org/10.1007/978-3-658-50113-6_11

Circle (Chap. 4; see also: Schiemann, 2006). According to Reichenbach (1938), the *context of discovery* is the context that can tell us something about the origin of scientific theories. Unlike the *context of justification,* the context in which we examine the empirical validity of scientific theories using a well-developed canon of methods (that is, a collection of established research methods), we apparently have hardly any heuristics or even sets of rules with which we can set out in search of empirically valid theories. Feyerabend has criticized this dichotomy in several works and pointed out that even the context of justification is never based solely on "objective" procedures, but, like the context of discovery, is closely linked to much that is subjective and incalculable (see, for example, Feyerabend, 1980, p. 272; 1984, pp. 165; 1988, p. 162; 1992, pp. 157 see also *Chap. 14*).

The ongoing debates (e.g., Lugg, 1991; Pinch, 1989; Pučiliauskaitė, 2022) about the anthology with the provocative title "Farewell to Reason" can quite well be read as confirmation of the thesis: Paul K. Feyerabend is not forgotten.

Since his death in 1994, more than 17,000 publications have appeared under the keyword "Paul Feyerabend," according to *Google Scholar*. These include numerous new editions of his own writings as well as the aforementioned posthumously published works ("Conquest of Abundance" from 1999 and the German translation titled "Vernichtung der Vielfalt" from 2005, as well as "Naturphilosophie" from 2009), discussions of his critique of critical rationalism (e.g., Anderson, 2019), appreciations of his anarchist epistemology (e.g., Niaz, 2020), anthologies with critical essays (e.g., Bschir & Shaw, 2021; Preston et al., 2000) and memoirs (e.g., Stadler & Fischer, 2006), excellent biographies (e.g., Oberheim, 2007) and much more. He is also occasionally mentioned by name in feminist debates (e.g., Koertge, 2013); though not always in a positive light (e.g., Bowles, 1984, p. 186). Feyerabend, albeit en passant, takes a position on questions that are also raised by today's postcolonial studies. His concern about the epistemic and political marginalization of indigenous cultures, or about the human and ecological costs of Westernization, already preoccupied him in the 1970s. In the rich literature of postcolonial science and technology studies, Feyerabend unfortunately appears only as "[…] figure in the critical reactions against positivist philosophies of science…", which, while not incorrect, does not reflect the depth and distinctiveness of his critique of violent imperialist policy (Brown & Kidd, 2016, p. 5).

In "Against Method" and later works, Feyerabend criticizes scientific chauvinism, which often resists the introduction of scientific alternatives. Against such chauvinism, he sets the epistemic potential of myths, fairy tales, legends, and occult rites, and advocates for "citizens' initiatives instead of epistemology" (Feyerabend, 1980, p. 37).

But can publicly elected commissions of laypeople actually determine, "[...] whether the theory of evolution is really well founded, as the biologists want us to believe, whether a 'good justification' in their sense settles the matter, and whether other views, such as the doctrine of *Genesis* should also be presented in schools"? (Feyerabend, 1980, p. 190; emphasis in the original). Should views such as those presented by creationists be taught in schools? As is well known, Feyerabend would also like to place decisions about this in the hands of lay commissions (Feyerabend, 1980, p. 190).

Creationism (lat. creare: to create) refers to the view that the entire cosmos, including all living beings, was at least created by a creator god (see also Graf & Lammers, 2011, pp. 9). It is a religious belief that (largely) rejects biological evolution and is found in all Abrahamic religions (Christianity, Islam, and Judaism). The independent U.S. polling institute Pew Research Center surveyed people in 20 countries around the world in 2019/2020 about their views on evolution and belief in creation. In Germany, 17% of respondents agreed with the creationist statement that humans and other living beings have existed in their present form since the beginning of time. The lowest agreement with this statement was expressed by Spaniards and Swedes (10 and 11%, respectively). The highest agreement came from Americans (32%), Brazilians (36%), and people from Malaysia (55%) (Pew Research Center, 2020). In the USA, it is primarily white evangelical Protestants who believe that humans and other living beings have existed in their present form since the beginning of the world. In the USA, there are at least three major museums where belief in creation is presented as true knowledge and set in opposition to the findings of evolutionary research: the Creation Museum in Petersburg, Kentucky, the Ark Encounter in Williamstown, also in Kentucky, and the Museum of the Bible in Washington DC. All three museums are well attended. They do not simply promote the religious knowledge of their visitors, but above all a conservative political and racist worldview (see Thomas, 2020).

Can laypeople decide whether the laws of gravity *do not* apply, whether the coronavirus pandemic is an invention of Bill Gates, whether human-caused climate change is merely an illusion, or whether antisemitism is a lie spread by white women and men? To answer these questions with "yes" would be cynical. Thus, it should come as no surprise that Feyerabend's early call for citizens' initiatives instead of epistemology or philosophy is viewed with skepticism today and by some authors. For example, Philipp Sarasin (2019, p. 6) writes that we now recognize how toxic the combination of hostility to science, esoteric speculation, and identity-politics-driven insistence on the autonomy and uniqueness of every "culture" is. Although Feyerabend did not cause all this, he is nevertheless an early symptom of such developments. Alexander Bogner also believes that Paul Feyerabend, with his "multiparadigmatism," paved the way for today's conspiracy theorists (Bogner, 2021, pp. 62). Homayun Sidky argues similarly, accusing Feyerabend of having, with his egalitarian and anarchistic approach, created an intellectual space for all sorts of pseudoscientific views, such as magic, astrology, mysticism, and creationism (Sidky, 2020, p. 65). And Thomas Zoglauer points out that Feyerabendian relativism reaches its limits when lateral thinkers, fundamentalists, or conspiracy theorists are granted equal access to the media (Zoglauer, 2021, p. 52).

The significance of the innovative influence of (non-scientific) citizens' initiatives on social, political, and scientific change in contemporary times is certainly undeniable, whether it concerns the internationally active *MeToo movement,* the international *Fridays for Future* movement, or the *Black Lives Matter movement* (BLM), which also has an impact across national borders. Such citizens' initiatives aim not only to change individual lifestyles but also to mobilize politics, science, and citizens against sexual assault and exploitation, for a livable, climate-just future, and against racism. For Pierre Rosanvallon, these initiatives embody "counter-democracy," through which the politically active civil society monitors the state and its institutions (Rosanvallon, 2017).

However, such citizens' initiatives are not without controversy, as evidenced, for example, by antisemitic resentments within BLM or certain anti-feminist and racist positions in the MeToo movement (e.g., Funkschmidt, 2020; Martini, 2020).

However, today's citizens' initiatives also include groups that use of right-wing populist or right-wing extremist staging methods (e.g., the anti-Islamic movement "Pax Europa," the PEGIDA movement, the Reichsbürger scene). The call for citizens' initiatives instead of epistemology is therefore an ambivalent appeal. Feyerabend would certainly describe some "citizens' initiatives" today as "cranks," just as he did earlier (Feyerabend, 1964, p. 305).

Feyerabend only hints at what he means by cranks. Bschir and Lohse (2022) refer to Shaw (2021) and state: *Cranks* are—in Feyerabend's understanding—actors who seek only to impose their own views and agendas and are not interested in questioning their opinions in light of other perspectives. These can be activists who immunize themselves against other viewpoints, or local experts who are unwilling to consider views that contradict their personal experiences. Even scientific experts who see themselves as above all doubt can be cranks.

In short: Feyerabend by no means assumes that myths, fairy tales, occult rites, or indeed creationist beliefs are inherently superior to rational conceptions of the world. Rather, he criticized scientific attacks, for example, against astrology, herbal medicine, or parapsychology, when the attackers argue without knowledge or understanding of astrology or parapsychology. In doing so, Feyerabend is not defending the supposed pseudosciences. For Feyerabend, myths, fairy tales, or occult rites can instead serve both as a stimulus for scientific discoveries and—as it were—the *acid test* for the rationality of scientific research. Absurd ideas create cognitive dissonance because what they proclaim appears to contradict the familiar or the authoritatively proclaimed.

In the summer of 1983, Feyerabend organized a series of panel discussions at the Swiss Federal Institute of Technology in Zurich, dealing with "borderline problems of the sciences," including a seminar on parapsychology. Among those invited were representatives of parapsychological thought (Eberhard Bauer, Hans Bender, and Piet Hein Hoebens). In his contribution to the seminar discussion, Feyerabend said, among other things, that researchers in astrology, parapsychology, and the like continually strive to show "that they are respectable scientists, and that they have followed the canon of the sciences exactly. But what is this core, and what rules does it contain?" (in: Feyerabend & Thomas, 1985, p. 385). According to Ian James Kidd (2018), Feyerabend did not intend to defend parapsychology or astrology with his contribution. Rather, he was concerned with criticizing scientists who fail to live up to their own high intellectual standards when they simply ignore dissenting views instead of engaging with them scientifically. Thus, myths, fairy tales, but also scientific rationality point to interpretive patterns (or social constructions) that belong to the tried and tested, traditional patterns within communities, societies, and cultures, by means of which people interpret their world, communicate about it, and define their social identity—that is, their belonging to relevant reference communities—in order ultimately to find their way in this world.

In his later works, Feyerabend distances himself from a cultural relativism[1], according to which cultures must be regarded as "more or less closed units with

[1] He apparently also grew tired in later years of "relativizing everything and anything." "Yes, I myself," Feyerabend wrote to Hans Peter Duerr in August 1991, "have long been fed up with this worldview relativism, the principles of which you can find scribbled on the walls of every university shithouse…" (Feyerabend, 1995b, p. 279).

their own criteria and procedures" (1995a, p. 205). Rather, cultures are interconnected and learn from one another. Every culture potentially contains all cultures within itself. Cultural particularities are not sacrosanct. "There is no 'culturally justified' oppression and no 'culturally justified' murder. There is only oppression and murder, and both should be treated as such, and, if necessary, with determination. However, this mutability of every culture also leads to the insight that we must open ourselves to change before we try to change others" (1995a, p. 205).

In any case, in his later years Feyerabend spoke out decisively against reforming cultures, tribes, or nations from a Western perspective and with ignorant aggressiveness "in order to reform them and adapt them to their (the Western, WF) ideas of a civilized life. Ever since people were discovered who did not belong to the Western cultural and civilizational sphere, it was almost considered a moral duty to bring them the truth—and by that was meant the prevailing ideology of the conquerors" (Feyerabend, 1992, pp. 52). Feyerabend's criticism of the ideology of the conquerors also targets the "humanitarian mindset" as part of the ideology of a society (ibid., p. 59). This may seem irritating at first glance, but in view of the many historical and current crimes committed in the name of humanity, it is quite understandable (in detail: Frindte, 2022).

The encounter between ancient thinkers and writers such as Cicero, Seneca, Virgil, and the early humanists—if one may already call them that—began in the 14th century. Francesco Petrarca (1304–1374), Poggio Braccolini (1380–1459), and others entered into a virtual dialogue with the ancient authors in order to educate themselves and to perfect human civilization, humanitas. The name that this education received, following Cicero, was "studia humanitatis." And soon, just as diligently as they reflected on *humanus,* the human, they also pondered *humanitatis,* humanity, and *humanitas,* being human. With the *studia humanitatis,* the early humanists—though they did not yet call themselves that—formulated an educational program aimed at promoting both linguistic-formal education and substantive knowledge through the appropriation of ancient authors, and in this way to develop a new person modeled after the ancients (Buck, 1970, p. 123).[2] According to Erich Fromm, to recall the great humanist, the core of the humanistic worldview is the belief in the unity of humanity, the emphasis on human

[2] The German term *Humanismus* is associated above all with the name of Friedrich Immanuel Niethammer (1808). However, as early as 1781, a book by Alexander von Gleichen-Russwurm appeared with the title "Kultur und Geist der Renaissance" (Culture and Spirit of the Renaissance) and the formative subtitle "Das Jahrhundert des europäischen Humanismus" (The Century of European Humanism) (von Gleichen-Russwurm, 1781).

dignity and the capacity for people to develop and perfect themselves, as well as the highlighting of reason, objectivity, and peace (Fromm, 1966, pp. 117–122). Whether Paul Feyerabend was familiar with Fromm's humanism cannot be determined. There is no evidence for this in the publications. However, it is likely that Feyerabend would not have been very satisfied with the terms and categories Fromm uses to characterize the humanistic worldview, since, in his opinion, categories lead to nothing at all—unless power is conferred upon them (Feyerabend, 2005, p. 294). This possibility of exercising power with categories and concepts is probably Feyerabend's main reason for rejecting humanitarian sentiment as part of ideology and for strongly opposing attempts to impose such ideologies on other cultures, tribes, or nations (Feyerabend, 1992, p. 59). Feyerabend's criticism is not directed against humanistic visions, but rather against the intolerance, power, and violence of "Western" cultures.[3]

References

Andersson, G. (2019). Karl Popper und seine Kritiker: Kuhn, Feyerabend und Lakatos. In G. Franco (ed.), *Handbuch Karl Popper* (717–731). Springer VS.

Bogner, A. (2021). *Die Epistemisierung des Politischen. Wie die Macht des Wissens die Demokratie gefährdet.* Reclam.

Bowles, G. (1984). The Uses of Hermeneutics for Feminist Scholarship. *Women's Studies International Forum, 7*(3), 185–188.

Brown, M. J., & Kidd, I. J. (2016). Introduction: Reappraising Paul Feyerabend. *Studies in History and Philosophy of Science Part A, 57,* 1–8.

Bschir, K., & Lohse, S. (2022). Pandemics, Policy, and Pluralism: A Feyerabend-inspired perspective on COVID-19. *Synthese, 200*(6), pp. 441.

Bschir, K. & Shaw, J. (2021). *Interpreting Feyerabend. Critical essays.* Cambridge University Press.

Buck, A. (1970). Italienischer Humanismus: Forschungsbericht. *Archiv für Kulturgeschichte, 52*(1), 121–140.

Feyerabend, P. K. (1951). *Zur Theorie der Basissätze.* Dissertation. Universität.

Feyerabend, P. K. (1964). Realism and Instrumentalism: Comments in the logic of factual support. In M. Bunge (ed.), *Critical approaches to science and philosophy* (pp. 260–308). The Free Press.

Feyerabend, P. K. (1980). *Erkenntnis für freie Menschen.* Suhrkamp.

Feyerabend, P. K. (1981a). *Realism, Rationalism and Scientific Method: Philosophical Papers Volume 1.* Cambridge University Press.

[3] In "Farewell to Reason," Feyerabend notes in passing that one thing should be clear: fascism is not his "cup of tea," despite his pronounced sentimentality and his almost instinctive inclination to act humanely (Feyerabend, 1987, p. 309).

Feyerabend, P. K. (1981b). *Problems of Empiricism: Philosophical Papers Volume 2.* Cambridge University Press.
Feyerabend, P. K. (1984). *Against Method.* (5th edn.) Thetford Press.
Feyerabend, P. K., & Thomas, C. (1985). Was heißt das, wissenschaftlich sein? In P. Feyerabend & C. Thomas (eds.), *Grenzprobleme der Wissenschaften* (pp. 385–397). Eidgenössische Technische Hochschule, Verlag der Fachvereine.
Feyerabend, P. K. (1987). *Farewell to reason.* Verso.
Feyerabend, P. K. (1988). Knowledge and the role of theories. *Philosophy of the Social Sciences, 18*(2), 157–178.
Feyerabend, P. K. (1992; Original: 1989). *Über Erkenntnis. Zwei Dialoge.* Campus.
Feyerabend, P. K. (1995a). *Zeitverschwendung.* Suhrkamp.
Feyerabend, P. K. (1995b). *Briefe an einen Freund, herausgegeben von Hans Peter Duerr.* Suhrkamp.
Feyerabend, P. K. (2005). *Die Vernichtung der Vielfalt. Ein Bericht, herausgegeben von Peter Engelmann.* Passagen.
Frindte, W. (2022). *Quo Vadis, Humanismus.* Springer.
Fromm, E. (1966). The global philosophy of Man.*The Humanist, Yellow Springs, Ohio, 26,* 117–122.
Funkschmidt, K. (2020). Der Antisemitismus in der „Black Lives Matter "-Bewegung und seine Ursprünge. *Materialdienst—Zeitschrift für Religions- und Weltanschauungsfragen, 83*(5), 358–366.
Graf, D., & Lammers, C. (2011). Evolution und Kreationismus in Europa. In D. Graf (ed.), *Evolutionstheorie—Akzeptanz und Vermittlung im europäischen Vergleich* (pp. 9–28). Springer Berlin.
Kidd, I.J. (2018). Feyerabend, Pluralism, and Parapsychology. *Bulletin of the Parapsychology Association, 10*(1), 5–9.
Koertge, N. (2013). Feyerabend, Feminism, and Philosophy. *HOPOS: The Journal of the International Society for the History of Philosophy of Science, 3*(1), 139–141.
Lugg, A. (1991). Reviewed Work: Farewell to Reason by Paul Feyerabend. *Canadian Journal of Philosophy, 21*(1), 109–120.
Martini, F. (2020). Wer ist# MeToo? Eine netzwerkanalytische Untersuchung (anti-) feministischen Protests auf Twitter. *M&K Medien & Kommunikationswissenschaft, 68*(3), 255–272.
Niaz, J. R. M. (2020). *Feyerabend's Epistemological Anarchism.* Springer Nature Switzerland.
Niethammer, F. P. I. (1808). *Der Streit des Philanthropinismus und Humanismus in der Theorie des Erziehungs-Unterrichts unsrer Zeit.* Friedrich Frommann.
Oberheim, E. (2007). *Feyerabend's Philosophy.* De Gruyter.
Pew Research Center. (2020). Biotechnology Research Viewed With Caution Globally, but Most Support Gene Editing for Babies to Treat Disease; https://www.pewresearch.org/science/2020/12/10/biotechnology-research-viewed-with-caution-globally-but-most-support-gene-editing-for-babies-to-treat-disease/. Accessed: March, 25, 2023.
Pinch, T. (1989). Reviewed Work: Farewell to Reason by Paul Feyerabend. *American Journal of Sociology, 94*(6), 1457–1459.
Preston, J., Munévar, G., & Lamb, D. (Eds.). (2000). *The worst enemy of science?—Essays in memory of Paul Feyerabend.* Oxford University Press.

Pučiliauskaitė, S. (2022). Paulio Feyerabendo socialinės minties kritinė apžvalga. *LOGOS-A Journal of Religion, Philosophy, Comparative Cultural Studies and Art, 110*, 17–23. https://www.ceeol.com/search/article-detail?id=1070970. Accessed: 10. Mai 2023.

Reichenbach, H. (1938). *Experience and prediction. An analysis of the foundations and the structure of knowledge.* The University of Chicago Press.

Rosanvallon, P. (2017). *Die Gegen-Demokratie. Politik im Zeitalter des Misstrauens.* Hamburger Edition.

Sarasin, P. (2019). *„Anything goes": Paul Feyerabend und die etwas andere Postmoderne.* Geschichte der Gegenwart. https://www.zora.uzh.ch/id/eprint/182292/. Accessed: March, 25, 2023.

Schiemann, G. (2006). Inductive Justification and Discovery. On Hans Reichenbach's Foundation of the Autonomy of the Philosophy of Science. In J. Schickore & F. Steinle (eds.), *Revisiting Discovery and Justification. Historical and Philosophical Perspectives on the Context Distinction* (pp. 23–39). Springer.

Shaw, J. (2021). Feyerabend and Manufactured Disagreement: Reflections on Expertise, Consensus, and Science Policy. *Synthese, 198*(25), 6053–6084.

Sidky, H. (2020). *Science and anthropology in a post-truth world: A critique of unreason and academic nonsense.* Lexington Books.

Stadler, F., & Fischer, K. R. (Hrsg.). (2006). *Paul Feyerabend. Ein Philosoph aus Wien.* Springer.

Thomas, P. (2020). *Storytelling the Bible at the Creation Museum, Ark Encounter, and Museum of the Bible.* T&T Clark, Bloomsbury Publishing.

Von Gleichen-Russwurm, A. (1781). *Kultur und Geist der Renaissance. Das Jahrhundert des europäischen Humanismus.* Hoffmann und Campe Verlag.

Zoglauer, T. (2021). *Konstruierte Wahrheiten: Wahrheit und Wissen im postfaktischen Zeitalter.* Springer Vieweg.

12 Feyerabend and Psychology—An Excursion with Speculations

"In psychology courses, one reads about experiments and theories, but not Turgenev. […] Psychologists, ecologists, experts in interpersonal relations can […] learn a great deal from poets, novelists, actors such as Aeschylus, Lessing, or Brecht…" (Feyerabend, 1992a, pp. 142).

That is certainly true. But what was Paul Feyerabend's stance on psychology? His view of psychology became more differentiated over the decades. Some of his remarks are surprisingly accurate, others rather superficial. I will limit myself to a few examples:

While studying physics and astronomy in Vienna, he also attended psychology courses with Hubert Rohracher in the late 1940s and early 1950s. It is said to have been six semesters (Kuby, Limbeck-Lilienau & Schorner, 2010, p. 408). Rohracher (1903–1972) had studied law in Innsbruck and psychology in Munich, earned doctorates in both fields, and qualified as a professor in 1932 with a thesis on the theory of the will. He then worked at the University of Innsbruck and at times in Vienna. There, he was in close contact with Karl and Charlotte Bühler as well as with Egon Brunswick and his later wife Else Frenkel, a collaborator of the Bühlers.

Charlotte and Karl Bühler emigrated to the USA in 1933. Egon Brunswick was appointed to a professorship at the University of Berkeley in 1936. Else Frenkel also emigrated to the USA in 1938 and married Egon Brunswick there. Alongside Daniel J. Levinson, R. Nevitt Sanford, and Theodor W. Adorno, she is one of the main authors of the 1950 volume "The

W. Frindte, *Against Narrow-Mindedness and Chauvinism—Journeying Through Absurd Times with Paul K. Feyerabend*,
https://doi.org/10.1007/978-3-658-50113-6_12

Authoritarian Personality." Since 1943—after becoming a naturalized US citizen—Theodor Wiesengrund-Adorno, as is well known, called himself only "Theodor W. Adorno." Stone et al., (1993, p. 13) therefore consider it an "irony of fate" that only because of Adorno's official name change in 1943, "The Authoritarian Personality" is not cited as "Frenkel-Brunswik et al."

In 1938, Rohracher was temporarily stripped of his teaching license for political reasons. During World War II, he worked as a military psychologist. In 1943, he was appointed associate professor in Vienna. In June 1947, he was made full professor at the University of Vienna. By that time, he had already made a name for himself, especially for his research on the relationship between brain physiology and psychological processes. In doing so, he primarily used the electroencephalographic experiments developed by Hans Berger (1924/1929 in Jena) (see in detail: Benetka, 1998; Sachse & Goller, 2020). Rohracher advocated a decidedly natural-scientific, experimental psychology. For him, the EEG was an important instrument for the "objective view into the psyche of others" (Guttmann, 2009, p. 85). Without wanting to reduce the psychological to neurophysiological processes, he saw in these processes the causes of conscious experiences. And he considered the investigation of precisely these "conscious processes and states" to be the task of psychology as a science (Rohracher, 1988, pp. 2; original: 1947). He rejected the "doctrine" of unconscious psychological processes, because such hypotheses had been rendered obsolete by advances in neurophysiological research (cited in Guttmann, 2009, p. 85).

It is by no means far-fetched to suppose that Rohracher's courses may have influenced Feyerabend's skepticism toward psychoanalysis as much as his later engagement with the mind–body problem, that is, the relationship between the human mind (the psychological) and the brain (the physical).

The mind–body problem (or the psychophysical problem) can safely be described as a fundamental problem of philosophy: What is the relationship between body (physical body, brain) and mind (consciousness, the psychological)? This question has troubled ancient thinkers (Plato, Aristotle, Leucippus, Democritus, Epicurus, and others) as much as the early Enlightenment philosophers (e.g., René Descartes, Gottfried Wilhelm Leibniz,

Christian Wolff, 1679–1754), and more recent philosophers, neuroscientists, and psychologists (e.g., David Chalmers, 2021; Hilary Putnam, 1994; Gerhard Roth, 2019; Wolf Singer, 2020, among many others) cannot let it go either. Do body and mind interact (as Descartes assumed)? Is the psychological a non-physical property of matter (as David Chalmers argues)? Are the soul, the mind, the psychological, products or functional principles of neurophysiological processes (as Gerhard Roth and Wolf Singer claim, and as was also advocated in the almost forgotten psychology of the GDR; see, e.g., Klix, 1971, p. 18)? A rather radical, but by no means absurd, view is held by proponents of social constructionism (Gergen, 1994).[1] They deliberately ignore the mind–body problem and claim: The concepts and forms through which we understand the world and ourselves are social products/constructions of a historically and culturally shaped process of exchange between individuals and groups. The mind–body problem, too, is a social construction. In order to perceive and recognize it as a problem at all, one must participate in cultural traditions in which the autonomy of mental processes is significant. That is, the mind–body problem is part of a language game in which one can participate. But one can also refrain and take part in another language game, in which, for example, the social construction of knowledge plays a greater role than the interaction of mind and body (Gergen, 2010).

In his frequently cited essay "Materialism and the Mind–Body Problem," Feyerabend concludes—after attempting to refute various arguments against materialism and briefly mentioning Gestalt psychology—that there is not a single reason

[1] The term *social constructionism* or *Social Constructionism* is used by Kenneth Gergen (1985) to distinguish, on the one hand, from constructivist conceptions of science that deal with individual processes of reality construction (as in radical constructivism; Von Glasersfeld, 1997), and, on the other hand, to establish a connection to social science approaches (e.g., Berger and Luckmann, 2000; original: 1966), which are primarily concerned with the social processes by which conceptions of reality are generated and tested for their usefulness.

why the attempt to provide a purely physiological explanation of human beings should be abandoned (Feyerabend, 1963, p. 65).[2] Some interpreters took Feyerabend's insistence on the "purely physiological account of human beings" as an occasion to accuse him of denying the independent existence of mental (psychic) processes. He is said to have advocated—at least in his earlier writings—an eliminative materialism, according to which there are only material and no mental entities (e.g., Everit, 1981; Bechtel & Hamilton, 2007). Jamie Shaw (2021), however, doubts that Feyerabend was an "eliminative materialist": "Feyerabend Never was an Eliminative Materialist." While Feyerabend did defend materialism (and naturalism) in his early works, for example in the 1963 essay, he later drew attention to a principle suitable for judging worldviews, theories, and even myths by whether they are meaningful and practical for shaping life. Shaw refers here to what Feyerabend, in "Vernichtung der Vielfalt" or in "Conquest of Abunddance", calls the "Aristotelian principle": "Real is that which plays a key role in the life with which we identify" (Feyerabend, 2005, p. 215; first published: 1992b).[3] This "principle" is really just another formulation of the social and scientific pluralism on which the "late" Feyerabend repeatedly insisted. Also in "Vernichtung der Vielfalt," Feyerabend addresses the mind-body problem. He argues that it remains relevant even though neurophysiology has provided detailed models of mental processes. Some scientists, on the other hand, demand that "mind and our consciousness" be granted again the influence they had before the rise of materialist psychology. That is, there are indeed people in science who resist the "[...] elimination of pre-scientific psychological ideas and entities" (2005, p. 153). Whether Feyerabend considers such resistance helpful, appropriate, compatible, or unscientific is not revealed at this point. However, one page later he writes that the natural sciences "[...] exclude 'subjective' elements such as feelings and sensations" (ibid., p. 154), "even though they play a role in their acquisition and control (of scientific research, WF)" (ibid.; original: 1989). This also means that all efforts in (natural) scientific research toward objectivity of knowledge are always linked, as Wolfgang Pauli (1954) puts it, to the subjective factor, that is, they cannot be separated from the subjective states of the observers. By the late 1980s, this

[2] "There is, therefore, not a single reason why the attempt to give a purely physiological account of human beings should be abandoned, or why physiologists should leave the 'soul' out of their considerations" (Feyerabend, 1963, p. 65).

[3] Or as Kurt Lewin once put it: "Real is what works" (Lewin, 1969, p. 41; original 1936).

was likely a commonplaçe even in the "objective" sciences. "The researchers," Feyerabend writes, "influence the material and are also influenced and changed by it, for after all, it is the material from which they themselves are formed. [...] The 'subjective' side of knowledge cannot simply be discarded, especially since it is inextricably intertwined with its material manifestations" (2005, p. 158).

If one interprets Feyerabend's statements that explicitly or implicitly refer to psychological matters in his later works (as I do, in any case), one might think that the "late" Feyerabend had distanced himself from his earlier, "materialist," but certainly reductionist, views on the nature of the mental and the uselessness of psychological theories. I will return to this point. However, his skepticism toward a "purely physiological" explanation of human beings and their psyche may have begun even earlier.

Between 1949 and 1953, the Austrian College financed Paul Feyerabend's first trips abroad. These took him to Germany (Göttingen), Denmark (Copenhagen, Askov), Sweden (Lund, Stockholm, Uppsala), and Norway (Ustaoset). In Askov, he met Niels Bohr, and in Copenhagen he met Edgar Tranekjaer-Rasmussen (1900–1994), whom he already knew from Alpbach and who probably also belonged to the guests of the "Kraft-Kreis" (Kuby, 2010, p. 5). Through Tranekjaer-Rasmussen, Feyerabend became acquainted with the phenomenological-psychological approach of Edgar Rubin (1886–1951) and its further development by Tranekjaer-Rasmussen himself. Not only is the "Rubin vase," the black-and-white reversible figure, named after Edgar Rubin. Rubin significantly influenced the work of (German) Gestalt theory on figure-ground processes, that is, the scientific studies of the mutual influence of foreground and background in the perception of arbitrary objects (Katz, 1951). This mutual influence of figure and ground in perception also points, as is well known, to the relativity of sensory impressions; people do not simply reproduce the world, but rather construct it for themselves, not least on the basis of internal (physiologically and/or ontogenetically acquired) structural laws (Gestalt laws; Bühler, Schlaich & Sinner, 2017; pp. 29–38).

Shortly before his death, Edgar Rubin published an essay entitled "Visual figures apparently incompatible with geometry" (Rubin, 1950), in which he reports, among other things, on experiments he conducted in part with Tranekjaer-Rasmussen. In these experiments, subjects were confronted with variations of the well-known "Müller-Lyer figure" (a geometric optical illusion). The relatively simple result initially shows the contradictions between the geometric representation of the figures and individual perceptions, namely the illusions. The "Müller-Lyer figure" shows two lines of equal length, whose ends are marked either (line a) by two arrowheads or (line b) by two arrow tails. In many cases, line b is perceived as longer than line a (see also *Chap. 14*). Moreover, the results also point to the relativity of individual sensory data. Today we know that optical illusions arise automatically and unconsciously during the perceptual process and only become accessible to consciousness afterward.

Paul Feyerabend was not only familiar with Rubin's work, but apparently also participated in some perception experiments under the guidance of Tranekjaer-Rasmussen (see, e.g., Feyerabend, 1978, p. 12, p. 57; original: 1958; 1995, p. 107). Daniel Kuby, Christoph Limbeck-Lilienau, and Michael Schorner (2010) now suspect that the research in perceptual psychology that Feyerabend encountered in Denmark also fostered his doubts about the "theory of sense data" and the foundations of logical empiricism. In "Against Method," Feyerabend emphasizes that the production of observational statements (or basic statements) consists of two psychological events: a clear and unambiguous perception *and* a clear and unambiguous connection between this perception and certain parts of language (Feyerabend, 1986, pp. 95). Perhaps, through the experiments he observed in Copenhagen, he already realized at that time that at least the first event—the unambiguous perception of objects and events—is not so easy to accomplish. Perceptions obey figure-ground relationships, and these relationships often change as a result of strong emotions, he will write elsewhere, at a later point (Feyerabend, 1987a, pp. 78).

Let us continue our stroll through Feyerabend's psychology: There is much talk of psychology, psychological claims, or psychological problems (e.g., Feyerabend, 1981). Very early on—according to a stamp in the book with Feyerabend's Vienna address at Alliogasse 14, as early as 1948—and thoroughly (judging by the many underlinings and annotations), he read the influential book "Perception and the World of Objects: Foundations of a Psychology from the Standpoint of the Object) by Egon Brunswick (Brunswick, 1934). The book is part of Feyerabend's personal library (Paul Feyerabend Archiv, University of Konstanz). Brunswick attempts, on the basis of logical empiricism, to develop a psychology of perception that, by means of mathematical and statistical methods, should be able to quantify perceptual judgments. This may have suited the "taste" of the early Feyerabend.

In his later works, references can be found to Wilhelm Wundt (1832–1920) and his *Völkerpsychologie* (e.g., Feyerabend, 2009, p. 83), to Jean Piaget (1896–1980) and the innate schemata of perception (e.g., Feyerabend, 1987a, p. 703; 2009, p. 154). He cites Jerome Bruner (Feyerabend, 1988, p. 170; see also below), has taken note of works by the Soviet neuropsychologist Alexander R. Luria (1902–1977) and the British neurologist Oliver Sacks (1933–2015) (Feyerabend, 2005, p. 27; Feyerabend, 1988, pp. 159), or mentions the well-known essay "The conflict between Aristotelian and Galilean Mode of Thought in contemporary Psychology" by Kurt Lewin (Feyerabend, 1978, p. 232; original: 1972). And of course, one might assume, he also thoroughly worked through "Die Psychologie des Aristoteles" (The Psychology of Aristotle) by Franz Brentano (1967; original: 1867) (Feyerabend's personal library).

As already mentioned, Feyerabend followed the work of perception psychology with particular curiosity; for example, the work of Julian Hochberg (1923–2022), who is among the internationally recognized perception psychologists and, among other things, dealt with the (automatic) processing of (moving) images (Feyerabend, 1984a, p. 96; Hochberg, 1983). I have already mentioned his interest in the perception-psychological work of Edgar Tranekjaer-Rasmussen. The references to the work of Gestalt psychologists David Katz (1884–1953)[4] and Oswald Külpe (1892–1915), the phenomenologist Maurice Merleau-Ponty (1908–1961), or the social psychologist Muzafer Sherif (1906–1988)[5] are yet to be added (e.g., Feyerabend, 1978, pp. 48, 143, 261; originals: 1960, 1966, 1967; Feyerabend, 1986, p. 19). He was also familiar with Leon Festinger's (1957) theory of cognitive dissonance and considered it a good explanation for dogmatism, as can be read in a Zurich lecture manuscript on epistemology from 1984 (Paul Feyerabend Archive, University of Konstanz, PF-4–7–1; there is also an offprint of an article by Leon Festinger, Henry Riecken, and Stanley Schachter, 1956, in which the important assumptions of the theory formulated a year later are already contained; PF-1–11–1).

Feyerabend was also impressed by the striking experiments that the Belgian psychologist Albert Michotte (1881–1965) had already conducted in the 1940s on the perception of causality.

Between 1905 and 1908, Albert Michotte worked as a visiting scholar in Leipzig with Wilhelm Wundt and in Würzburg with the Gestalt psychologist Oswald Külpe. In 1944, he founded an institute for psychology at the University of Leuven. There, he experimented with a setup also known as the "Banc Michotte." In this setup, participants were asked to judge mechanically moved paper discs

[4] David Katz emigrated to England in 1933 and to Sweden in 1937. There he received a professorship in psychology at the University of Stockholm.

[5] The Turkish-American social psychologist Muzafer Sherif conducted numerous social-psychological experiments on the emergence of norms in groups (Sherif, 1966; original: 1936) and, among other things, developed the theory of realistic group conflict (Sherif, 2015; original: 1967), to which the Polish-British social psychologist Henri Tajfel would later also refer (see below).

on which lines were drawn. The participants saw only a section of the discs and were asked to say what they saw. Depending on the presentation, they would say, for example: Object A gave B a shock, chased it away, or threw it out. That is, the participants described in their responses seemingly perceived causal relationships (cf. Leyssen, 2021). One could also call it "phenomenal causality." It is a subjective attribution of causes and effects. In social psychology, Fritz Heider (1896–1988) experimented in the 1940s with similar experimental setups and later developed the associated early theory of attributions (Heider, 1944).

In any case, Feyerabend was enthusiastic about Michotte's experiments, seeing in the "[...] epoch-making investigations of phenomenological causality" a confirmation that the "[...] laws of perceptual organization [...] are not completely determined by the available information..." (Feyerabend, 1981, p. 156). Perception psychology was therefore important to him, as he regarded it as an outstanding tool with which historians of science and philosophy should also be familiar, as can be read in "Über Erkenntnis" (Feyerabend, 1992a, p. 102). A similar remark can be found a few years earlier: "A fundamental knowledge of the psychology of perception must, I believe, sooner or later lead to the insight that the idea of the indubitability of sensations is not only a logical but above all a psychological myth" (Feyerabend, 1978, p. 58; original: 1960). This is also directed against empiricism in the natural sciences, namely against the "old" theory of sense data. The results of our perception, whether in the observation of physical experiments or in everyday self- and other-observation, are never unambiguous "sense data" and cannot be. They are individual constructions against the background of our embedding in various social contexts. In this respect, it is not so absurd to claim that "[...] it is not the existence of sense data that regulates behavior in situations of perception, but quite the reverse, the compulsion to behave in certain ways guarantees the existence of sense data" (Feyerabend, 1978, p. 50).

That would be one side. On the other hand, in psychology (and medicine), the incursion of "abstract traditions" has led to the elimination of many important and useful results as well as to the accumulation of interesting, "objective," but for practical purposes completely useless knowledge, as stated in a footnote in "Knowledge for Free People" (1980, p. 117). Here, Feyerabend is probably also referring to the work of Paul Meehl and Hans-Jürgen Eysenck (ibid.); both influential, much-cited psychologists inclined toward critical rationalism and critics of psychoanalysis (e.g., Eysenck, 2004; original: 1985; Meehl, 1973). Feyerabend discussed the mind-body problem and the relationship between theory and experiment with Paul Meehl at the end of the 1970s and the beginning of the 1980s (Feyerabend, 1995, p. 161).

Paul Everett Meehl (1920–2003) was a psychologist and professor of psychology at the University of Minnesota, as well as a former president of the American Psychological Association. He was one of the most productive psychologists of the 20th century. The range of his work includes contributions to schizophrenia, statistical methods in psychology, and the relationship between theory and experiment. Together with Herbert Feigl and Wilfrid Sellers, he founded the Minnesota Center for the Philosophy of Science, which Feyerabend also frequently visited. In 1954, Meehl published a book on clinical and statistical predictions that sparked an intense controversy. Meehl argued that statistically-automated prediction methods, when used correctly, enable more efficient and reliable decisions regarding the prognosis and treatment of patients than individual diagnoses. His conclusions were controversial and for a long time contradicted the prevailing consensus on psychiatric decision-making. Among Paul Feyerabend's papers are, among other things, some offprints of Paul Meehl's works from the years up to 1971 (Paul Feyerabend Archive, PF 8–2; PF 8–3). Meehl's book "Selected philosophical and methodological papers" (Meehl, 1991) is also found in Feyerabend's personal library.

"Scientific psychology" has eliminated the soul (Feyerabend 1980, p. 139). By "scientific psychology," he presumably means the behaviorist, cognitivist, empirically-experimental research paradigms that shaped psychology from the 1950s to the 1980s, and thus certainly also the work of Paul Meehl. Psychologists of the humanistic tradition, among whom he counts, for example, the psychoanalyst Erik H. Erikson (1902–1994), are the praiseworthy exception, because they have preserved their openness of perspective and a natural humanity (Feyerabend, 1980, p. 158). Psychoanalysis, which reintroduced the idea of personal (psychic) causes of mental illness (Feyerabend, 1984a, p. 167), is a blessing especially for those thinkers who like to talk about the "nature of man" but do not subject themselves to the effort of detailed study (Feyerabend, 2009, p. 78). This is not directed against Erikson, but specifically against Sigmund Freud's "The Interpretation of Dreams" (1912) and Carl Gustav Jung's works on the developmental history of thought (1912). It seems as if Feyerabend's brief dismissal of Freud's "Interpretation of Dreams" does not exactly speak of great familiarity. He had expressed himself similarly years earlier in a small footnote about Freud. Psychoanalysis, with the symbolism it uses, has introduced a pictorial language and interpreted it in such a way that it is inconceivable how it could be refuted (Feyerabend, 1955, p. 465).

A first edition of "The Interpretation of Dreams" appeared as early as November 1899 and was then postdated to 1900. In this book, Freud presents his discoveries about the *unconscious* (Freud, 1987; original: 1900). The "unconscious" was not unknown in philosophy, Romanticism, and early psychology. But with his discovery, Freud not only laid the foundation for the understanding and therapy of psychological suffering, primarily neurotic disorders; he also proposed a differentiated order of psychic structures. The psychic cannot be reduced to the conscious; rather, the unconscious is the truly psychic. However, a distinction must be made between the unconscious, which is incapable of consciousness, and a preconscious, which can become conscious (Freud, 1987, p. 499). So, in the end, we are dealing with three levels: the unconscious, the preconscious, and the conscious. The unconscious comprises the repressed contents of inadequately processed experiences and events. Freud's discovery of the stratification of the psyche and the influence of the unconscious on body and mind is likely one of the great moments in psychology. The psychic, our human mind, refers not only to the past, present, and future, but also to the hidden, repressed, or forgotten from past, present, and future. A similar, contemporary formulation can be found in John Bargh (2017), who is credited with anchoring the scientific study of the unconscious in academic psychology at the end of the 20th century (see *Chap. 14*). Feyerabend could not possibly have known this. His, let us say it plainly, narrow-minded view of Sigmund Freud and psychoanalysis may be clouded by ignorance, prejudice, and arrogance. The spirit of the times may also have played a role, even if Feyerabend would vehemently resist the suggestion that he ever followed the spirit of the times.

While the members of the first Vienna Circle saw in Freudian psychoanalysis "approaches to deeper explanation" to account for the "misguided paths of metaphysics" (Verein Ernst Mach, 2006, p. 13; original: 1929), skepticism and rejection of psychoanalysis developed in "scientific" psychology despite institutional expansion already in the first decades of the 20th century. During the National Socialist era, this grew into defamation, in which "established" scientists also participated. For example, Eduard Spranger (1882–1963), the flamboyant prolific writer and inventor of the "Third Humanism," claimed that psychoanalysis destroyed the "spiritual health of the people" (Spranger, 1933; quoted in Ortmeyer, 2008, p. 34). At the end of 1933, psychoanalysts in Germany were informed that psychoanalysis would only be tolerated if all Jewish members were excluded from the German Society for Psychoanalysis. And soon after, the expulsions began. Sigmund Freud left Austria in June 1938. The non-Jewish staff of the famous *Berlin Psychoanalytic Institute* were integrated in 1936 into the National Socialist "German Institute for Psychological Research and Psychotherapy," which was headed by Matthias Heinrich Göring, a cousin of Hermann Göring (see also Blum, 1999). After 1945, psychoanalysis did experience a new boom in Germany, Austria, and the USA, but this took place mainly outside of "academic" psychology. Else Frenkel-Brunswick, herself a trained psychoanalyst, wrote in 1940 in American exile: "The conceptual apparatus of psychoanalysis is not sufficient to fully explain rational and social behavior" (Frenkel-Brunswick, quoted in Sprung, p. 243). Perhaps this brief statement captures the spirit of the times that still surrounds psychoanalysis today, and of which Paul Feyerabend also caught a whiff.

When discussing Paul Feyerabend's understanding of psychology, his references to Benjamin Lee Whorf (1897–1941) should not be overlooked. Whorf was not a psychologist, but he nevertheless significantly influenced Feyerabend's ideas about the relationship between worldview, thinking about the world, and linguistic structures (e.g., 1986, pp. 295; 2005, p. 50; 2009, p. 93). And this relationship is indeed an eminently psychological one. The so-called Sapir-Whorf hypothesis—Edward Sapir was Whorf's teacher—states, as is well known, that an individual's perception of the world and thinking about the world are strongly determined by the (native) language of the perceiving individual.[6] Languages, as Feyerabend writes in "Against Method," are not merely means of describing events, but they also constitute events. Language encompasses stories about the world, society, and people. These stories influence people's thinking, behavior, and perception (Feyerabend, 1986, p. 295). The Sapir-Whorf hypothesis was controversial for a long time. Extreme forms, according to which perception and thinking about the world are *completely* determined by the respective language, can hardly be maintained empirically. But that language influences our thinking about the world and—possibly—also the constitution of cultural practices is undisputed; one need only consider, for example, the role language plays in the construction of gender, sexuality, or race (e.g., Dong, 2022; see also: *Chap. 15*).

[6] "The categories and types we isolate from the world of phenomena are not found there because they stare each observer in the face; rather, the world presents itself in a kaleidoscopic flow of impressions that must be organized by our minds—and that means largely by the linguistic systems of our minds" (Whorf, 1940, p. 213).

The influence of Ernst Mach (1838–1916) should also definitely be mentioned, as Paul Feyerabend engaged with him intensively and repeatedly (Stadler, 2006, p. XXV). As is well known, not only the members of the "Vienna Circle" were influenced by Ernst Mach; his work (and that of Christian Freiherr von Ehrenfels) on sensory physiology also forms part of the foundation upon which later Gestalt psychology was built. For example, Mach discovered, among other things, the Mach Band Effect named after him, an optical illusion in the perception of different brightness levels. Mach's efforts to overcome psychophysical dualism (in which, simply put, brain and mind are regarded as different entities) and to formulate a doctrine "[…] as an empirical unity of physics, physiology, and psychology within the framework of a monistic worldview" (Stadler, 2019, p. 2; Mach 1872), likely found fertile ground with Feyerabend. The insight that psychic (mental) processes and structures cannot be investigated by a single method is also something Feyerabend owes to his reading of Mach's works. According to Mach, the "mental field," that is, the realm of thoughts, emotions, and aspirations, cannot be explored solely through self-observation, but must be supplemented, for example, by neurophysiological methods (Feyerabend, 1984b, p. 10).

Extended digression: With his empirio-criticism, Ernst Mach advocated what we would today call a constructivist theory about the relationship between reality and sensations. While, to briefly look at the history of psychology, for Theodor Fechner and Wilhelm Wundt a sensation follows a stimulus, for Mach sensations are not results, "[…] but the basic building blocks, the elements from which the world itself arises" (Benetka & Slunecko, 2019 pp. 105). At the beginning of the 20th century, Ernst Mach's philosophical views became especially popular in Russia. Some of Lenin's associates considered how to bring the ideas of socialism closer to the "simple" Russian people. The philosopher and economist *Alexander Alexandrovich Bogdanov* (1873–1928), the cultural politician and later People's Commissar for Education *Anatoly Vasilyevich Lunacharsky* (1875–1933), and other Russian intellectuals sympathized in this context with the philosophical views of the Austrian physicist and philosopher. The sympathizers tried to reconcile Mach's ideas with the views of the Russian Social Democratic Labour Party. Lenin did not like this at all. He saw in Mach's empirio-criticism a "[…] veritable campaign against the philosophy of Marxism" (Lenin, 1975, vol. 14, p. 9). He, in turn, wanted to go on the offensive. This was also about a demonstration of power against the renegades from his own ranks. Although his elementary philosophical knowledge should actually have precluded such a demonstration of power, Lenin, as the physicist Friedrich Adler wrote to Ernst Mach, "... really crammed all the literature, but did not have the time to think it through" (Adler, letter to Mach of June 23, 1909; quoted in Wittich, 1999, pp. 82). Perceptions and ideas, Lenin claimed, are images of things that exist outside of us (Lenin, 1975, vol. 14, p. 103); human thought "is then 'economical' when it *correctly* reflects the objective world" (ibid., p. 166; emphasis in the original). "To recognize objective truth, i.e., truth independent of man and of mankind, means to recognize absolute truth in one way or another" (ibid., p. 127). One could now criticize Lenin's image theory from today's admittedly somewhat condescending perspective. For example, one could claim: Sensations are not only images of stimuli from the external world, but also the result of the brain's own dynamic neurophysiological processes. Human thinking can also be economical when it relies on schemas and heuristics that may provide a simplified representation of the objective world and yet somehow manage to cope with it. More productive than such modern-day bashing, however, are comparisons between Lenin's image theory and the state of knowledge at the time. One could recall Immanuel Kant and the laws of nature that the human mind *ascribes* to it; or the unconscious, through which—according to Freud—humans involuntarily take a stance toward reality and the objective external world; or the findings of the Würzburg school of thought psychology, according to which people are subjects of their own thinking *(Chapters 12, 14)*. In short: With the image theory and the fight against Ernst Mach's empiriocriticism, Lenin fell behind those insights that could certainly be counted among the human-scientific achievements regarding sensations, thinking, reason, and reality up to that point. Paul Feyerabend, who liked to irritate his critics with references to Lenin, occasionally writes that Lenin was as familiar with the natural science of his time as few other authors. In a footnote, however, Feyerabend adds that he "[...] is thinking here mainly of Lenin's remarks on Hegel's 'Logic' and 'History of Philosophy.' It is different with 'Materialism and Empirio-criticism'" (Feyerabend, 1981, p. 265). And a few pages later, Feyerabend again refers to Lenin, who criticizes a crude materialism and indeed attributes a positive function to idealism in the history of ideas. However, Lenin did not apply this principle of discovering something positive in idealism to Mach. For Mach, "[...] with his resistance to the dominance of mechanics, paved the way for a more dialectical conception" (Feyerabend, ibid., p. 268). This is likely also directed against the mechanistic image theory in the Leninist mold.

As is well known, not only neurophysiological but also other experimental and qualitative methods are part of the inventory of pluralistic research in contemporary psychology. And it is also not particularly noteworthy, since it is now largely known, though by no means unanimously accepted, that such methodological pluralism is underpinned by a highly differentiated psychological science in terms of disciplines, paradigms, theories, and applications, whose representatives, however, do not always see eye to eye. Feyerabend's example likely points to the lesser evil: "Divergences proliferate in psychology: behaviorists and neurophysiologists despise introspection which is an important source of knowledge for gestalt psychology, clinical psychologists rely on their experience, sometimes called 'intuition', i.e. on the reaction of their own well prepared organism, while more 'objective' schools use rigorously formulated test instead" (Feyerabend, 1987b, p. 75).

In other words: By the end of the 1980s, Paul Feyerabend—at least to some extent—was not only aware of the process of differentiation within psychology as a science[7]; he also perceived the disputes and narrow-minded attempts within and between psychological communities to either ignore or combat the diversity of methods and theories. However, his perception was not particularly profound, considering that by the late 1970s and early 1990s, psychology as a whole had to cope with turbulent changes: Psychology had undergone a "cognitive turn," with behaviorism being replaced in the 1960s and 1970s. As early as the late 1940s, Bruner and Postman experimentally demonstrated that and how needs, motivations, and expectations (or "mental states") influence perception (Bruner & Postman 1949). Traditional Gestalt psychology was merged with cognitive psychology (Flores d'Arcais, 1975). Kenneth J. Gergen criticized "cognitivism" in psychology because it could not explain the social relationships in which cognitive structures arise (Gergen, 1985, 1989). In the summer of 1971, Philip

[7] In his view that a process of differentiation was taking place in psychology and that various theoretical and methodological approaches were now competing, Feyerabend likely saw himself confirmed by Jerome Bruner's autobiography (Feyerabend, 1988, p. 170; Bruner, 1983).

Zimbardo and colleagues—not far from Feyerabend's place of work in Berkeley—conducted their famous, controversial, and yet so revealing *Stanford Prison Experiment* (Haney, Banks & Zimbardo, 1973). In 1967, Henri (Hers Mordche) Tajfel, born in Poland, was appointed professor in Bristol, England, and initiated a "social turn" in social psychology (Tajfel & Turner, 1979). After the unrest in Paris in May 1968, Serge Moscovici reflected on social change brought about by minorities (1979). Edward T. Hall and Mildred R. Hall endeavored to understand differences and similarities between cultures (e.g., Hall & Hall, 1987). In Berlin and elsewhere, Critical Psychology was established, with Klaus Holzkamp and colleagues seeking to position psychology as a means of critiquing capitalist power relations (see: *Chap. 14*). The people being studied were to be included in research as equal subjects and learn to become independently capable of action (e.g., Holzkamp, 1983; Markard, 2018). And so on and so forth.

It is possible that Paul Feyerabend may have followed similar discussions.[8] In the manuscript "Naturphilosophie" (written in the early 1970s, published posthumously in 2009), there is a brief reference to the "[…] investigations of modern psychology into the development of self-consciousness […]" (Feyerabend, 2009, p. 323). It is conceivable that Feyerabend may well have observed those developments in psychology in which "[…] with the emergence of countercultural alternative movements after 1968, the idea of the authentic self" (Amlinger & Nachtwey, 2022, p. 19) became the starting point for theoretical conceptualizations. One might think, for example, of impression management theory (Tedeschi et al., 1971), the theory of self-efficacy (Bandura, 1977), or the theory of symbolic self-completion (Wicklund & Gollwitzer, 1981), among others.

In fact, during his time in Zurich, Paul Feyerabend engaged with topics that are also very important to me. In preparation for lectures at the Swiss Federal Institute of Technology (ETH) in Zurich from 1984 to 1990, he dealt, among other things, with the postmodern ideas of Jean-François Lyotard and the social criticism of Joseph Weizenbaum (Paul Feyerabend Archive, University of Konstanz, PF 4–16–1)[9]. He compared the political and intellectual conditions in Plato's time with contemporary problems and drew attention to global dangers

[8] For example, Feyerabend's personal library includes the anthology "Physics of Cognitive Processes: Amalfi 1986," edited by Eduardo R. Caianiello (1987). In one of the contributions, Edward T. Hall presents his cultural dimensions, and Feyerabend diligently underlines numerous passages (Paul Feyerabend Archiv, University of Konstanz).

[9] The book "Postmodernism for Children" (Lyotard, 1987; original: 1986) is in Feyerabend's personal library (Paul Feyerabend Archiv, University of Konstanz).

such as nuclear war or worldwide ecological catastrophes (Paul Feyerabend Archive, PF 6–1–10).

Together with his assistant Christian Thomas, he prepared a lecture series entitled "In Search of the Real," which was held in the summer semester of 1990 at ETH. In their correspondence, Feyerabend and Thomas developed the plan for this lecture series, which also concerned the funding of the project, in order to have it finally approved by the dean of their department (Paul Feyerabend Archive, PF 4–5). A focus of this lecture series was the constructivist approaches to the conception of reality by Paul Watzlawick and other authors. As suggestions for the lecture, Feyerabend proposed, among others, the books "Wie wirklich ist die Wirklichkeit" (How Real Is Real?) (Watzlawick, 1985; original: 1976) and "Die erfundene Wirklichkeit – Wie wissen wir, was wir zu wissen glauben" (Invented Reality—How Do We Know What We Think We Know? Contributions to Constructivism) (Watzlawick, 1985; original: 1981). Both books are in Feyerabend's personal library and are heavily annotated and underlined. The second book, in particular, is revealing. In addition to Paul Watzlawick, who comments on self-fulfilling prophecies, prominent representatives of radical constructivism are included here: Ernst von Glaserfeld, Heinz von Foerster, and Francisco Varela.

It is certainly not speculative to claim that the "late" Paul Feyerabend sympathized with constructivism. In a note for a lecture he gave in the summer semester of 1985 at ETH Zurich, there is, among other things, a sentence that also points to Feyerabend's interest in constructivism: The idea of an "objective" reality is widespread, but is slowly being replaced by constructivism (Paul Feyerabend Archive, University of Konstanz, PF 6–2–2). Compared to his early views from the 1950s and 1960s, this would indeed represent an epistemological turn. More than that: It is not constructivism as such—which does not seem to exist as a monolithic theoretical edifice (Mosayebi, 2003)—that interests Paul Feyerabend, but rather—so I argue—the relationships, conflicts, and interactions between individual and social constructions of reality. Among other things, Feyerabend's personal library contains the book "Denken und Sprechen" (Thought and Language) by Lev S. Wygotski (1986; original: 1934). Before I address Feyerabend's annotations in this book, I must offer some explanations: Lev Semionovich Vygotsky (1896–1934) is usually cited as the founder of the Soviet *Cultural-Historical School* (together with Alexander Romanovich Luria and Alexei Nikolaevich Leontiev), although he neither founded an institute nor a scientific school, and his collaboration with Leontiev and Luria was limited in time and by no means free of conflict (Friedrich, 1991). As is well known, Vygotsky was interested not only in the psychology of art (the subject of his dissertation, Wygotski, 1976; original: 1925), but above all in the individual development of psychological processes in general and in the relationship between speech and thought in particular. A topic, therefore, that also preoccupied Paul Feyerabend. Vygotsky assumed a) that psychological contents and structures (knowledge, attitudes, thought processes, etc.) are of social origin, b) that they must first be appropriated and internalized by the individual, and c) that this process of appropriation, interiorization, is an interactive, cooperative (or, as it is called today, a collaborative) process in which learners and teachers jointly construct reality. Vygotsky's approach can safely be described as social constructivist (Boyland, 2019). And so it is not surprising that the adherents of social constructivism or constructionism have intensely received Vygotsky's works—at the latest since they began to appear in English in the 1980s—and count him, so to speak, among their intellectual forebears (see, e.g., Gergen, 1999; Harré, 2002; Shorter, 1995).

The bouquet of "fathers" to whom social constructivists refer is diverse: Aristotle, Baruch de Spinoza, Giambattista Vico, Immanuel Kant, Friedrich Nietzsche, Ludwig Wittgenstein, Ludwik Fleck, Thomas Kuhn, Alexander Romanovich Luria, George Herbert Mead, Michel Foucault, as well as Feyerabend and many others are frequently invoked (e.g., Burr, 2015; Gergen, 1999; Shotter, 1993). In social constructivism, there are at least two conceptions of the psychological. The first I would call a more moderate variant, as represented for example by Gün Semin, in which the psychological is recognized as a relevant scientific problem area. Following Vygotsky, for Semin all higher mental processes are the result of internalized social relations (Semin, 1990, p. 162). A society is only possible to the extent that interacting "selves" share the same basic symbolic system. That is: for Semin, psychological and social processes are in an interactive relationship, in which the socio-historical forms of human interaction exercise a dominant function. In the second, more radical variant of social constructionism, represented above all by Kenneth Gergen, the psychological as a psychological object plays no or only a very subordinate role. All feeling, thinking, and acting of the individual is the result of social constructions as negotiated in the process of social communication. "In effect, what we label as individualized characteristics—aggressivness, playfulness, altruism, and the like—are primarily products of joint configurations" (Gergen & Gergen, 1988, p. 41). This statement, that aggressiveness, altruism, etc. are products of a joint, social process, is, however, ambiguous. It initially refers only to the socially conditioned labels, not yet to the underlying socially conditioned psychological processes themselves. In a German-language article, we find a clearer position: "In modernist times, the fully developed self resembled a well-made product, a unity that persisted independently of time and circumstances. In contrast, with postmodernity, the individual ceases to exist as an independent, autonomous unit. If individuals are the result of relationships, then one must conclude that relationships are more fundamental than individuals" (Gergen, 1990, p. 197). Thus, while in the more moderate version of social constructivism the psychological is still conceived of and treated as a subject for psychological research, in the more radical version it figures only as a label meant to designate something else.

Now, after these somewhat cumbersome excursions, to return to the aforementioned book by Vygotsky: Feyerabend underlined the following (famous) sentence there and annotated it with "Arist!" (for Aristotle): "The development of the child's thinking does not proceed from the individual to the socialized, but from the social to the individual" (Wygotski, 1986, p. 44). This is truly a social-constructivist statement (in the aforementioned moderate variant). Did Feyerabend sympathize with this moderate version of social constructivism? And why did the sentence remind him of Aristotle? John Preston believes that in his later works Feyerabend abandoned the core of scientific realism and embraced a social-constructivist metaphysics (Preston, 1997, pp. 426). Preston refers, among other things, to Feyerabend's work "Art as a Product of Nature as a Work of Art" (Feyerabend, 1994). The sentence that Preston cites, among others, to support his arguments, is also found in "Die Vernichtung der Vielfalt." There, Feyerabend writes that he will argue that "[...] our entire universe, from the mythical Big Bang through the emergence of hydrogen and helium, of galaxies, fixed stars, planetary systems, viruses, bacteria, fleas, and dogs up to the glorious appearance of Western man, is an *artifact*, constructed by generations of *scientific craftsmen* from a material of unknown properties that sometimes yields, sometimes resists" (Feyerabend, 2005, p. 240; emphasis in the original). Now, one can say that Feyerabend does not deny the existence of the universe, the Big Bang, or fleas, but rather assumes that our knowledge and our discourse about the universe, etc., are the result of (social) constructions created by individuals and communities and—following Vygotsky—can be appropriated. The world—and this is a central assumption of social constructionism—is constructed in social exchange and can only be perceived and understood through the socially created, that is, constructed, concepts and constructions. These concepts and constructions can be appropriated by us humans and examined according to the criteria of their suitability for further social exchange. And so perhaps Feyerabend's annotation "Arist!" to the above-mentioned sentence by Vygotsky can be explained. Possibly, so my speculation, Feyerabend was reminded of the already mentioned *Aristotelian principle,* which I would like to quote once again: "Real is that which plays a key role in the life with which we identify ourselves" (Feyerabend, 2005, p. 215). This principle also explains, according to Feyerabend, "why so many different processes (visions, immediate experiences, dreams, and religious fantasies) have been declared real and why discussions about reality arouse so much passion" (Feyerabend, ibid.).

Whether these are sufficient reasons to discover a proponent of social constructivism in the "late" Feyerabend, I do not know. I would certainly find it very appealing.

References

Adorno, T. W.; Frenkel-Brunswick, Else Levinson, Daniel J., & Sanford, R. Nevitt (1950). *The Authoritarian Personality*. Harper & Row.

Amlinger, C., & Nachtwey, O. (2022). *Gekränkte Freiheit. Aspekte des Libertären Autoritarismus*. Suhrkamp.

Bandura, A. (1977). Self-efficacy: Toward a unifying theory of behavioral change. *Psychological Review, 84*, 191–215.

Bargh, J. (2017). Before you know it: The unconscious reasons we do what we do. Simon & Schuster.

Bechtel, W., & Hamilton, A. (2007). Reduction, Integration, and the Unity of Science: Natural, Behavioral, and Social Sciences and the Humanities. In T. A. F. Kuipers (Eds.), *General Philosophy of Science: Focal Issues* (pp. 377–430). Elsevier.

Benetka, G., & Slunecko, T. (2019). Ernst Machs Bedeutung für die Herausbildung einer naturwissenschaftlichen Psychologie—Zur Geschichte eines Missverständnisses. In F. Stadler (Eds.), *Ernst Mach—Zu Leben, Werk und Wirkung* (pp. 99–109). Springer Nature Switzerland.

Benetka, G. (1998). Entnazifizierung und verhinderte Rückkehr: Zur personellen Situation der akademischen Psychologie in Österreich nach 1945. *Österreichische Zeitschrift für Geschichtswissenschaften, 9*(2), 188–217.

Berger, P., & Luckmann, T. (2000; Original: 1966). *Die gesellschaftliche Konstruktion der Wirklichkeit: Eine Theorie der Wissenssoziologie*. Fischer.

Blum, H. P. (1999). Die Vertreibung der Psychoanalyse und die Wiederkehr des Verdrängten. *Forum Psychoanalyse, 15*, 348–359.

Boyland, J. R. (2019). A social constructivist approach to the gathering of empirical data. *Australian Counselling Research Journal, 13*(2), 30–34.
Brentano, F. (1967; Original: 1867). *Die Psychologie des Aristoteles, insbesondere seine Lehre vom nous poietikos*. Wissenschaftliche Buchgesellschaft.
Bruner, J. S., & Postman, L. (1949). Perception, cognition and behaviour. *Journal of Personality, 18*, 14–31.
Bruner, J. S. (1983). *In Search of Mind: Essays in Autobiography*. Harper & Row.
Brunswick, E. (1934). *Wahrnehmung und Gegenstandswelt: Grundlegung einer Psychologie vom Gegenstand her.* Deuticke.
Bühler, P., Schlaich, P. & Sinner, D. (2017). *Visuelle Kommunikation: Wahrnehmung—Perspektive—Gestaltung*. Springer Vieweg.
Burr, V. (2015). Social Constructionism. *International Encyclopedia of the Social & Behavioral Sciences*, 2nd edition, Volume 22. Routledge.
Caianiello, E. R. (1987). *Physics of Cognitive Processes: Amalfi 1986*. World Scientific Publications.
Chalmers, D. (2021). Idealism and the mind-body problem. In J. Farris & B. P. Göcke (Eds.), *The Routledge Handbook of Idealism and Immaterialism* (591–613). Routledge.
Dong, J. (2022). A study on the relationship between language and thought based on Sapir-Whorf hypothesis. *Journal of Global Humanities and Social Sciences, 3*(3), 83–85.
Everitt, N. (1981). A problem for the eliminative materialist. *Mind, 90*(359), 428–434.
Eysenck, H. (2004). *Decline and Fall of the Freudian Empire*. Routledge.
Festinger, L. (1957). *A theory of cognitive dissonance*. Stanford University Press.
Festinger, L., Riecken, H. W., & Schachter, S. (Original: 1956). *When prophecy fails. Experiencing social psychology: Readings and projects (69–75)*. Minneapolis: University of Minnesota Press.
Feyerabend, P. K. (1955). Wittgenstein's Philosophical Investigations. *The Philosophical Review, 64*(3), 449–483.
Feyerabend, P. K. (1958). An attempt at a realistic interpretation of experience. *Proceedings of the Aristotelian Society*, Vol. 58, 143–170).
Feyerabend, P. K. (1960). Das Problem der Existenz theoretischer Entitäten. In E. Topitsch (Eds.), *Probleme der Wissenschaftstheorie. Festschrift für Viktor Kraft*. Springer Verlag.
Feyerabend, P. K. (1963). Materialism and the mind-body problem. *The Review of Metaphysics, 17*(1), 49–66.
Feyerabend, P. K. (1978). *Der wissenschaftstheoretische Realismus und die Autorität der Wissenschaften, ausgewählte Schriften*. Vieweg & Sohn.
Feyerabend, P. K. (1980; Original: 1978). *Erkenntnis für freie Menschen*. Suhrkamp.
Feyerabend, P. K. (1981). *Probleme des Empirismus*. Vieweg & Sohn.
Feyerabend, P. K. (1981; Original: 1966). *Probleme des Empirismus*. Braunschweig/Wiesbaden: Vieweg & Sohn.
Feyerabend, P. K. (1984a). *Wissenschaft als Kunst*. Suhrkamp.
Feyerabend, P. K. (1984b). Mach's theory of research and its relation to Einstein. *Studies in History and Philosophy of Science Part A, 15*(1), 1–22.
Feyerabend, P. K. (1986; Original: 1975). *Wider den Methodenzwang*. Suhrkamp.
Feyerabend, P. K. (1987a). Creativity: A Dangerous Myth. *Critical Inquiry, 13*(4), 700–711.

Feyerabend, P. K. (1987b). *Farewell to Reason*. Verso.

Feyerabend, P. K. (1988). Knowledge and the Role of Theories. *Philosophy of the Social Sciences, 18*(2), 157–178.

Feyerabend, P. K. (1989a). Realism and the Historicity of Knowledge. *The Journal of Philosophy, 86*(8), 393–406.

Feyerabend, P. K. (1992a; Original: 1989). *Über Erkenntnis. Zwei Dialoge*. Campus.

Feyerabend, P. K. (1992b). Historical Comments on Realism. In A. van der Merve, F. Selleri, & G. Tarozzi (Eds.), *International Conference on Bell's Theorem and the Foundations of Modern Physics* (194–202). World Scientific Pub.

Feyerabend, P. K. (1994). Art as a Product of Nature as a Work of Art. *World Future, 40*, 87–100.

Feyerabend, P. K. (1995). *Zeitverschwendung*. Suhrkamp.

Feyerabend, P. K. (1999; Original: 1960). The Problem of the Existence of Theoretical Entities. In P. K. Feyerabend, *Knowledge, Science and Relativism*, Philosophical Papers (Vol. 3), herausgegeben von John Preston (pp. 16ff.). Cambridge University Press.

Feyerabend, P. K. (2005). *Die Vernichtung der Vielfalt. Ein Bericht*, herausgegeben von Peter Engelmann. Passagen Verlag.

Feyerabend, P. K. (2009). *Naturphilosophi*e, herausgegeben von H. Heit und E. Oberheim. Suhrkamp.

Flores d'Arcais, G.B. (1975). Einflüsse der Gestalttheorie auf die moderne kognitive Psychologie. In S. Ertel, L. Kemmler, & M. Stadler (Eds.), *Gestalttheorie in der Modernen Psychologie*. Steinkopff.

Freud, S. (1987). *Original: 1900*. Die Traumdeutung: Fischer.

Friedrich, J. (1991). Die Legende einer einheitlichen kulturhistorischen Schule in der sowjetischen Psychologie. *Deutsche Zeitschrift für Philosophie, 39*(5), 536–546.

Gergen, K. J. & Gergen, M. M. (1988). Narratives and The Self as Relationship. In: L. Berkowitz (Ed.). *Advances in Experimental Social Psychology*. Vol. 21. Academic Press.

Gergen, K. J. (1985). The social constructionist movement in modern psychology. *American Psychologist, 40*(3), 266–275.

Gergen, K. J. (1989). Social psychology and the wrong revolution. *European Journal of Social Psychology, 19*(5), 463–484.

Gergen, K. J. (1994). *Realities and Relationships. Soundings in Social Construction*. Harvard University Press.

Gergen, K. J. (1999). *An Invitation to Social Construction*. London: Sage (dt. 2002: Konstruierte Wirklichkeiten. Eine Hinführung zum sozialen Konstruktionismus. Kohlhammer.

Gergen, K. J. (2010). The acculturated brain. *Theory & Psychology, 20*(6), 795–816.

Gergen, K. J. (1990). Die Konstruktion des Selbst im Zeitalter der Postmoderne. *Psychologische Rundschau., 41*, 191–199.

Guttmann, G. (2009). Die Neurowissenschaften auf der Suche nach dem Unbewussten. In P. Giampieri-Deutsch (Eds.), *Geist, Gehirn, Verhalten. Sigmund Freud und die modernen Wissenschaften*. Königshausen & Neumann.

Hall, E. T., & Hall, M. R. (1987). *Hidden differences: Doing business with the Japanese*. Anchor Press/Doubleday.

Haney, C., Banks, W. C., & Zimbardo, P. G. (1973). A study of prisoners and guards in a simulated prison. *Naval research reviews*, *9*(1–17).
Harré, R. (2002). Public Sources of the Personal Mind. *Theory & Psychology, 12*(5), 611–623.
Heider, F. (1944). Social perception and phenomenal causality. *Psychological Review, 51*(6), 358–374.
Hochberg, J. (1983). Problems of Picture Perception. *Visual Arts Research, 9*(2), 7–24.
Holzkamp, K. (1983). *Grundlegung der Psychologie*. Campus.
Jung, C. G. (1912). Wandlungen und Symbole der Libido. Beiträge zur Entwicklungsgeschichte des Denkens. Fr. Deutieke.
Katz, D. (1951). Edgar Rubin. 1886–1951. *Psychological Review, 58*(6), 387–388.
Klix, F. (1971). *Information und Verhalten*. Verlag der Wissenschaften.
Kuby, D. (2010). Paul Feyerabend in Wien 1946–1955. Das Österreichische College und der Kraft Kreis. In Benedikt, M., Knoll, R., Schwediauer, F. & Zehetner, C. (Eds.), *Auf der Suche nach authentischem Philosophieren. Philosophie in Österreich 1951–2000. Verdrängter Humanismus -verzögerte Aufklärung*. (1041–1056). Facultas. wuv.
Kuby, D., Limbeck-Lilienau, C. & Schorner, M. (2010). Paul Feyerabend—Die Dogmen des Logischen Empirismus. Editorische Notiz. In F. Stadler (Eds.), *Vertreibung, Transformation und Rückkehr der Wissenschaftstheorie. Am Beispiel von Rudolf Carnap und Wolfgang Stegmüller* (pp. 403–409). LIT.
Lewin, K. (1969). *Grundzüge der topologischen Psychologie*. Huber.
Leyssen, S. (2021). Remaking "Michotte": Reusing and Remaking Moving Images in the History of Perception Research. *Isis, 112*(2), 315–325.
Lyotard, J.-F. (1987). *Original: 1986)*. Postmoderne für Kinder: Passagen.
Mach, E. (1872). *Die Geschichte und Wurzel des Satzes von der Erhaltung der Arbeit*. Calvesche K & K Universitätsbuchhandlung.
Markard, M. (2018). Kritische Psychologie. In O. Decker (Eds.), *Sozialpsychologie und Sozialtheorie* (pp. 107–121). Springer VS.
Meehl, P. E. (1973). Some methodological reflections on the difficulties of psychoanalytic research. *Psychological Issues, 8*, 104–115 (Original: 1970).
Meehl, P. E. (1991). *Selected philosophical and methodological papers*. Minneapolis, Oxford: University of Minnesota Press.
Moscovici, S. (1979). *Sozialer Wandel durch Minoritäten*. Urban & Schwarzenberg.
Ortmeyer, B. (2008). *Eduard Spranger und die NS-Zeit. Forschungsbericht 7.1*. Johann Wolfgang Goethe-Universität.
Paul Feyerabend Archiv, Universität Konstanz. https://www.kim.uni-konstanz.
Pauli, W. (1954). Wahrscheinlichkeit und Physik. *Dialectica, 8*(2), 112–124.
Preston, J. (1997). Feyerabend's retreat from realism. *Philosophy of Science, 64*(S4), S421–S431.
Putnam, H. (1994). Sense, nonsense, and the senses: An inquiry into the powers of the human mind. *The Journal of Philosophy, 91*(9), 445–517.
Rohracher, H. (1988; Original: 1947). *Einführung in die Psychologie*. Urban & Schwarzenberg.
Roth, G. (2019). *Aus Sicht des Gehirns*. Suhrkamp.

Rubin, E. (1950). Visual figures apparently incompatible with geometry. *Acta Psychologica, 7*, 365–387.

Sachse, P., & Goller, P. (2020). 6 Innsbruck—Kurz gefasste Geschichte des Instituts für Psychologie an der Universität Innsbruck (1896–1993/2000). In A. Stock & W. Schneider (Eds.), *(2020), Die ersten Institute für Psychologie im deutschsprachigen Raum: Ihre Geschichte von der Entstehung bis zur Gegenwart* (201–232). Hogrefe.

Semin, G.R. (1990). Everyday assumptions, language and personality. In K. J. Gergen & G. R. Semin (Eds.). *Everyday understanding*. Sage.

Shaw, J. (2021). Feyerabend Never Was an Eliminative Materialist. In K. Bschir & J. Shaw (Eds.), *Interpreting Feyerabend: Critical Essays* (114–131). Cambridge University Press.

Sherif, M. (1966). *Original: 1936)*. The Psychology of Social Norms: Harper & Row.

Sherif, M. (2015). *Original: 1967)*. Group Conflict and Co-Operation: Psychology Press.

Shorter, J. (1995). In dialogue: Social constructionism and radical constructivism. In L. P. Steffe & J. Gale (Eds.), *Constructivism in Education* (41–56). Routledge.

Shotter, J. (1993). *Cultural politics of everyday life: Social constructionism, rhetoric and knowing of the third kind*. University of Toronto Press.

Singer, W. (2020). *Ein neues Menschenbild?* Gespräche über Hirnforschung: Suhrkamp.

Sprung, H. (2011). Else Frenkel-Brunswick: Wanderin zwischen der Psychologie, der Psychoanalyse und dem Logischen Empirismus. In S. Volkmann-Raue & H. Lück (Eds.), *Bedeutende Psychologinnen des 20. Jahrhunderts.* (pp. 235–246). VS Verlag.

Stadler, F. (2006). Paul Feyerabend—Ein Philosoph aus Wien. In F. Stadler & K. R. Fischer (Eds.), Paul Feyerabend. Ein Philosoph aus Wien (pp. IX–XXX). Springer.

Stadler, F. (2019). Nur ein philosophischer „Sonntagsjäger"?—Der Naturforscher Ernst Mach. In F. Stadler (Eds.), *Ernst Mach—Zu Leben, Werk und Wirkung* (pp. 1–20). Springer Nature Switzerland.

Stone, W. F., Lederer, G., & Christie, R. (Eds.). (1993). *Strength and weakness*. The authoritarian personality today: Springer-Verlag.

Tajfel, H., & Turner, J. C. (1979). An integrative theory of intergroup conflict. In W. G. Austin & S. Worchel (Eds.), *The social psychology of intergroup relations* (pp. 33–47). Brooks/Cole.

Tedeschi, J. T., Schlenker, B. R., & Bonoma, T. V. (1971). Cognitive dissonance: Private ratiocination or public spectacle? *American Psychologist, 26*(8), 685.

Verein Ernst Mach (2006, Original: 1929). *Wiener Kreis. Texte zur wissenschaftlichen Weltauffassung von Rudolf Carnap, Otto Neurath, Moritz Schlick, Philipp Frank, Hans Hahn, Karl Menger, Edgar Zilsel und Gustav Bergmann*, herausgegeben von Michael Stöltzner & Thomas Uebel. Felix Meiner Verlag.

Von Glasersfeld, E. (1997). *Radikaler Konstruktivismus: Ideen, Ergebnisse, Probleme*. Suhrkamp.

Watzlawick, P. (1985; Original: 1976). *Wie wirklich ist die Wirklichkeit*. Piper.

Watzlawick, P. (Eds.) (1985; Original: 1981). *Die erfundene Wirklichkeit—Wie wissen wir, was wir zu wissen glauben? Beiträge zum Konstruktivismus*. Piper.

Whorf, B. L. (1940). Science and Linguistics. *Technology Review, 42*(2), 229–231.

Wicklund, R. A., & Gollwitzer, P. M. (1981). Symbolic self-completion, attempted influence, and self-deprecation. *Basic and Applied Social Psychology, 2*(2), 89–114.

Wittich, D. (1999). Lenins „Materialismus und Empiriokritizismus“—Entstehung, Wirkung, Kritik. *Sitzungsberichte der Leibniz-Sozietät*, *30*(3).
Wygotski, L. S. (1976; Original: 1925). *Die Psychologie der Kunst*. Verlag der Kunst.
Wygotski, L. S. (1986; Original: 1934, erste deutsche Veröffentlichung 1964). *Denken und Sprechen*. Fischer.

13 On the Absurd, Chauvinism, and Narrow-mindedness

"No thought is so old or absurd that it could not improve our knowledge" (Feyerabend, 1986, p. 55).

13.1 Feyerabend and the Absurd

Feyerabend often speaks of the absurd and absurdities. The above quote comes from "Against Method," and fifty pages later he asks how Galileo manages to introduce *absurd* and induction-defying claims (Feyerabend, 1986, p. 105). That we live in an "absurd world" is revealed in "Über Erkenntnis. Zwei Dialoge" (On Knowledge) (1992, p. 137). In his autobiography "Killing Time," he comments on Brecht's "Arturo Ui," which he did not like. The humor was crude, the plot *absurd*, and had no connection "[…] to the actual absurdity of that time" (1995a, p. 182). Feyerabend also occasionally finds "absurd statements" among certain philosophers who oppose Niels Bohr (Feyerabend, 1968, p. 310). Observations, ideas, or theories can be "absurd," as can be read in various places in the collected volume "Der wissenschaftliche Realismus und die Autorität der Wissenschaften" (Scientific Realism and the Authority of Science) (Feyerabend, 1978). And Feyerabend not infrequently argues using the logical figure of reductio ad absurdum (Feyerabend, 1986, p. 381; 1981, p. 25). And so on.

The *reductio ad absurdum* is a form of argumentation or an indirect proof by which an assumption can be refuted by showing that it would lead to a contradiction (an absurdity). It is a logical proof technique already mastered by the ancient Greeks. The first to argue with this figure of reasoning are said to have been Hippocrates of Chios and Zeno (Brückler, 2017, p. 155), and it is also found "in the

W. Frindte, *Against Narrow-Mindedness and Chauvinism—Journeying Through Absurd Times with Paul K. Feyerabend*,
https://doi.org/10.1007/978-3-658-50113-6_13

much-maligned (and never read) Aristotle…" (Feyerabend, 1986, p. 381). The reductio ad absurdum can also be understood as a form of immanent criticism of a text or claim, in which the text or claim is not confronted with another view, but the contradictions or inconsistencies within the text or claim itself are revealed. This strategy of criticism is the one Feyerabend uses most frequently (see also Hoyningen-Huene & Oberheim, 2006, p. 18).

However, it is not the words or forms of argument that matter, but the ideas Feyerabend introduced into the world and why he does so. He formulated seemingly absurd ideas that appear to contradict Western critical rationality. "There are," says Feyerabend, "events (obviously senseless actions, dreams, and so on) which, when considered in isolation, appear absurd, but point to reasons that are different from their overt appearance" (Feyerabend, 2005, p. 188). One could dismiss Feyerabend's ideas as preposterous, insane, foolish, or simply absurd. Thus, he not only discusses the *historical* role of astrology and Johannes Kepler's (demonstrable) fondness for writing horoscopes,[1] but also tries to discover a meaning in the (early) astrological ideas about the connection between solar activity, planetary constellations, and their effects on people, a meaning that could be quite important for "primitive" peoples. Hopi medicine, acupuncture, medieval witchcraft, various forms of herbal medicine, the idea that comets could portend wars,[2] and of course rain dances, are for Feyerabend not just fairy tales or long-disproven myths. Above all, they are traditions whose equal right to exist Feyerabend strongly advocated. "Science is a source of knowledge, indeed a very important source—but myths, fairy tales, tragedies, epics, and many other creations of non-scientific traditions are as well" (Feyerabend, 1992, p. 116).

One of Feyerabend's often-cited and much-criticized absurd ideas is precisely the claim that there is no physical theory that contradicts the idea that rain dances bring rain (Feyerabend, 1980, p. 129; 1992, pp. 97). This idea has not

[1] Kepler did not only create horoscopes for the military leader Wallenstein. His estate is said to contain over 1,170 horoscopes for around 900 people (Boockmann et al., 2008, p. 53).

[2] The observation of striking celestial events in order to "recognize" upcoming events, crises, and catastrophes early was common practice among court astrologers, priests, and city chroniclers in the 15th and 16th centuries. For example, Martin Luther saw in Halley's Comet of 1531 and in the appearance of a meteor in 1532 signs of the impending end of the world (Rohr, 2013, p. 375).

only been frequently circulated in newspaper features (e.g., Saltzwedel, 1999). In scientific publications, such ideas are either considered problematic (e.g., Hagner, 2017, p. 373), their obscure background is criticized (e.g., Bammé, 2006, p. 281), or they are interpreted as a reference to the plurality of cultural traditions (e.g., Zoglauer, 2021, p. 55). However, Feyerabend does not claim that rain dances are as reliable as computer-based weather forecasts or that rain can be conjured up by means of rain dances. Of course, the idea of rain dances runs counter to the vast majority of the sciences. But it is not enough for us to claim that rain dances are ineffective. "A rain dance must be performed with the proper preparation and under the proper circumstances, and these circumstances include the old tribal organizations and the associated mental attitudes" (Feyerabend, 1992, p. 96). One can regard this sentence as a typically Feyerabendian provocation. But it says more. By referring to the old tribal organizations and mental attitudes, Feyerabend alludes to the traditions and conventions of those people for whom it makes sense that rain dances bring rain. It is therefore a *tribal- or community-based* causality, which from the perspective of "Western" rationality is absurd. And this tribal- or community-based causality is part of what would today be called indigenous knowledge (also: Native Science). Indigenous knowledge is the result of a long and creative engagement with nature and, as traditional knowledge, can be a crucial instrument not only for preserving the cultural identity of indigenous peoples, but also for better understanding the relationship between humans and nature.

In this sense, and only in this sense, it would be narrow-minded or chauvinistic to claim that rain dances, myths, rituals, and fairy tales are not important identity-forming traditions in some cultures and subcultures. With the seemingly absurd thesis, Feyerabend points on the one hand to the absurdity of "Western" dominant cultures. Their ignorance, narrow-mindedness, and chauvinism toward the diversity of worldviews, philosophies, and cultural values is, so to speak, one side of the absurd world. On the other hand, it is precisely the seemingly absurd ideas, such as those of rain dances, myths, and fairy tales, that are, from Feyerabend's perspective, desirable in order to resist the narrow-mindedness of "Western" dominant cultures and to arrive at new, transgressive insights. Feyerabend had already drawn attention to the significance of absurd ideas in an early publication. In the essay "How to be a good empiricist" from 1963, he writes, among other things, that absurd is what runs counter to our established linguistic habits. The feeling of absurdity, however, indicates that the new, seemingly absurd idea is suitable either for criticizing old ideas and theories or for leading to revolutionary ideas. Such absurdities and conflicts of meaning are therefore desirable (Feyerabend, 1999a, p. 101; original: 1963).[3]

With his seemingly absurd ideas, Feyerabend sought to expose the absurdities of this world, the ignorance of scientific "experts," the disregard for the traditions of disadvantaged cultures, cultural injustices, and the chauvinism of certain ideologies and worldviews. One could also say that Feyerabend tries to fight the absurd with the absurd in this world. "I believe," he wrote to Hans Peter Duerr in 1974, "my very last 'standpoint,' which I will ultimately stick to, will be Dadaism: *This crazy life has no meaning and apparently one can only construct one by adding a great deal of new nonsense to the old nonsense. All one can do is tell fairy tales and thus temporarily entertain oneself and others" (Feyerabend,* 1995b, *p. 24; emphasis in original).*

Absurd ideas can have an effect. They create cognitive dissonances because what they proclaim appears to contradict what is familiar or authoritatively declared. Their effect can perhaps be compared to good jokes. Good jokes and Feyerabend's absurd ideas, by unexpectedly and almost *around the corner* questioning the *normal* and placing it in a different, not necessarily fitting context, draw attention to the fact that there seems to be more than what is called the *normal.* Or put differently: Feyerabend's ideas, like good jokes, point to patterns of interpretation (or social constructions) that have not yet been tried out.

If one now tries to make sense of the concept of the absurd as Feyerabend uses it, the following might be surmised: The world is absurd when the diversity of worldviews is not tolerated. Against this absurdity, seemingly absurd ideas, language games, or actions can be deployed. They challenge established habits of interpretation, language, and action. They point to what else might be possible beyond (the absurd) reality. Seemingly absurd events, ideas, statements, or actions can create cognitive dissonances and social conflicts, which can be endured through a change of perspective (one might also call it divergent thinking) as well as through a social and scientific pluralism (which does not aim for consensus, but enables dissent).

[3] "[...] we deem absurd what goes counter to well-established linguistic habits. The absence, from a newly introduced set of ideas, of synonymy relations connecting it with parts of the accepted point of view; the feeling of absurdity therefore indicate that the new ideas are fit for the purpose of criticism, i.e. that they are fit for either leading to a strong *confirmation* of the earlier theories, or else to a very revolutionary *discovery:* absence of synonymy, clash of meanings, absurdity are desirable" (here quoted from: Feyerabend, 1999a, p. 101; original: 1963).

The term *divergent thinking* was introduced into psychology by Joy Paul Guilford (1967). Divergent thinking is lateral thinking in the best sense of the word. While convergent thinking seeks a clearly defined solution to a problem, divergent thinking is the creative, open, and playful engagement with a variety of problem solutions, in which mental blocks and the premature exclusion of (even seemingly alternative) possibilities are to be avoided.

Absurd events, ideas, statements, or actions are usually considered absurd from the perspective of individuals, groups, or communities for whom rationality is set as the only possible criterion for the world. While for some—individuals, groups, or communities—the absurd events, ideas, statements, or actions are part of their non-absurd reality, other people, groups, and communities often react to the absurd with incomprehension, ignorance, (scientific) chauvinism, narrow-mindedness, intolerance, or violence. When incomprehension, rejection, chauvinism, or violence become dominant habits of interpretation, language, and action, it seems justified to speak of an absurd world.

13.2 Digressions—Camus and the Absurd

At this point, a digression is permitted to bring Albert Camus, the philosopher of the absurd, into play for a moment. Not because Paul Feyerabend was an existentialist or a moralist. Perhaps he was that too. In his early years, he read Søren Kierkegaard, the pathfinder of existentialism (Feyerabend, 1995a, p. 82) and later referred to him to point to the richness of human experience (e.g., Feyerabend, 1986, p. 25; 1980, p. 48; 2009, p. 314; see also: Kidd, 2011). Researchers have also claimed to find affinities with Jean-Paul Sartre in Feyerabend's works (e.g., Neto, 1991, p. 555). No, that is not the reason why Albert Camus cannot be overlooked here. Both Feyerabend and Camus were rebels and anarchists. Freedom was important to them, as was the diversity of ways of mastering life. And this in a world that appeared absurd to them.

We humans know that we are mortal. Is it not absurd, that is, nonsensical, contrary to reason, to know of one's own mortality and yet strive to find meaning in life? In the knowledge of our mortality, no rational arguments help to understand this mortality. Should we then not end our lives right away? This, as is well known, is the starting point for Albert Camus to reflect in "The Myth of

Sisyphus—An Essay on the Absurd" on the one truly serious philosophical problem, suicide, and to seek answers to the fundamental question of philosophy, "... whether life is worth living or not" (Camus, 1997, p. 10; original: 1942). It is about the existential question of the meaning of life, the answer to which threatens to founder on the chaos and meaninglessness of the world. However, it is not the world itself that is absurd. What is absurd is the relationship of meaning-seeking humans to the seemingly meaningless world. The absurd is "... that divorce between the longing mind and the disappointing world..." (ibid., p. 56).

Humans cannot escape death, Camus knows that. And yet, life is worthwhile in its everydayness, routines, surprises, adversities, injuries, and absurdities. For Camus, to live life with dignity and love means to revolt against the absurdity of the world. In Sisyphus, Camus sees the model, the "hero of the absurd." Sisyphus, "... the powerless and rebellious proletarian of the gods" (ibid., p. 126), rose up against the gods and was banished to the underworld for it. There, the stone was already waiting, which he had to laboriously roll up the mountain, only for it to roll back down again and again, so that it had to be brought up once more in a seemingly pointless manner. That is the fate of Sisyphus. He accepts it and scorns it by making the stone his own, the everydayness of his actions.

"That is the whole silent joy of Sisyphus. His fate belongs to him. His rock is his thing ... The struggle itself toward the heights is enough to fill a man's heart. We must imagine Sisyphus happy" (ibid., pp. 127).

Whether Camus himself was a happy man is another question.[4] He derives three conclusions from the absurd: his revolt, his freedom, and his passions (Camus, 1997, p. 69). These could also be the conclusions with which Paul Feyerabend would sympathize.

"I rebel, therefore we are," wrote Camus a few years after "The Myth of Sisyphus" in the essay collection "The Rebel" (2011, p. 39; original: 1951). With this, he expands his three conclusions. Camus is concerned not only with the revolt of the individual, their freedom, and passions. Revolt presupposes "a common bond, the solidarity of the chain, which makes people alike and allies them" (ibid.,

[4] In a diary entry from 1936, Camus notes: "What I wish for now is not to be happy, but only to be conscious. One imagines oneself to be separated from the world, but it is enough for an olive tree to rise up in the golden dust, it is enough for a few patches of beach to flash in the morning sun, for one to feel this resistance within oneself melt away. So it is with me. I become aware of the possibilities for which I am responsible. Every minute of life carries within it its value as a miracle and its face of eternal youth" (quoted in Reif, 1999, p. 171).

p. 367) and elevates this solidarity to the criterion of its success. Revolt is not the revolutionary overthrow that, as in the two great failed "revolutions" of the 20th century (National Socialism and Communism), either tries to break the common bond through terror and concentration camps (ibid., p. 321) or proclaims the creation of the "total human" (ibid., p. 320) as a future task that can only be solved by way of injustice, violence, and torture. Revolt is indignation and defiance in multiple senses: the absurdity of the relationship between meaning-seeking humans and the seemingly meaningless world cannot be abolished. But revolt is possible. The individual revolts in their *everyday actions here and now* against the absurdity of death; the "freedom he demands, he demands for all; the one he rejects, he forbids to all" (Camus, 2011, p. 371; emphasis in the original). As a social movement, revolt presupposes the renunciation of violence as a principle (ibid., p. 374) and promises no *eternal* justice, no historically necessary salvation narrative. Rather, everything is possible (ibid., p. 379). Revolt cannot do without a "peculiar love," without that solidarity which arises from the fundamental equality of people; it "is thus love and fruitfulness, or it is nothing" (ibid., p. 397).

"At the zenith of thought, the rebel thus rejects divinity in order to share in common struggles and a common fate. We choose Ithaca, the faithful earth, bold and sober thought, clear action, the generosity of the knowing human. In the light, the world remains our first and last love" (Camus, 2011, p. 399).

As is well known, many things can be compared. Apples and pears, fascism and communism, existentialism and philosophy of science, Paul Feyerabend and Albert Camus. The Camus experts may bear with me. To compare does not mean to equate the objects, persons, ideologies, or sciences being compared. To compare means to weigh against each other in order to find differences or similarities. Comparisons change perspective and, under good circumstances, can help find the door out of one's own echo chamber. And so I discover in Camus and Feyerabend—despite the probably significant philosophical differences (e.g., how and why they like to refer extensively to the Greeks, flirt with mythical stories, or engage with Nietzsche[5])—similar rebellious ambitions. Their conceptions of

[5] Feyerabend often refers to Nietzsche, for example, to emphasize the significance of myths as forms of explanation and representation of the world, thereby supporting his critique of established positions in dominant philosophies of science (e.g., 1984, p. 51; see also Heit, 2016). He may also have been fascinated by the "solitary outsider" (see also Preston, 2002). For Camus, Nietzsche—whom he evidently greatly admired—provides, among other things, the framework for criticizing nihilism and replacing it with an absolute affirmation of the world (Camus, 1951; 1997, pp. 69, 2011, pp. 95; see also Piper, 2002).

the absurd may differ. For Feyerabend, the world is absurd when the diversity of worldviews is not tolerated; for Camus, as mentioned, the absurd is the conflict between the longing spirit and the disappointing world. Accordingly, Feyerabend fights against the destruction of diversity with sometimes equally absurd means and ideas. We should not overlook this ambiguity of the absurd in Feyerabend.

Camus, on the other hand, calls for both the acceptance of human absurdity and, at the same time, for revolt against absurdity through a life in the here and now, in dignity and love, and by renouncing violence.

In the consequences that Feyerabend and Camus derive from dealing with the absurd world, there do seem to be some overlaps: both rebel against historical necessities and advocate for a world of relativity (Camus, 2011, p. 379; e.g., Feyerabend, 1980, pp. 17 ff.). They emphasize the freedom of the individual, a freedom that, however, finds its limits where others are humiliated or destroyed (Camus, 2011, p. 371; e.g., Feyerabend, 1980, pp. 295 f.). The freedom of others is the limit for the freedom of the individual. The balance of the human and nature is also important to both (Camus, 2011, p. 249; e.g., Feyerabend, 1984, p. 12), and pseudo-reason is abhorrent to them (Camus, 2011, p. 289; e.g., Feyerabend, 1986, pp. 397). Tolerance, solidarity, revolt against narrow-mindedness, as well as humanity and love of humankind, on the other hand, are for Albert Camus and Paul Feyerabend those high values that they seek to deploy against resentment, totalitarian ideologies, dogmatism, and nihilism (e.g., Camus, 2011, p. 397; 1950; Feyerabend, 1980, p. 19, 1984, p. 13).

Camus, writes Jan Christoph Suntrup, "does not belong to the species of intellectuals who 'abuse the fame they have acquired in science or art to leave their domain and criticize society and the existing order in the name of a global, dogmatic (vague or precise, moral or Marxist) conception of humanity,' as Sartre puts it, but rather to the species of ironists who are aware of the relativity of their own convictions. To invoke dialogue against all those 'who are convinced they are absolutely right' also means to intervene without assuming the modern role of the ideological legislator" (Suntrup, 2009, pp. 193).

This probably also applies to Paul Feyerabend. There is probably nothing more to say here about the commonalities between Albert Camus and Paul Feyerabend.

13.3 Chauvinism and Narrow-mindedness

Paul Feyerabend opposes a *scientific* chauvinism, when individual scientists or scientific communities consider their rules, norms, traditions, and supposed truths to be the best and reject and combat alternative rules and traditions of other communities (e.g., Feyerabend, 1986, p. 55; original: 1975). It may be that Feyerabend initially focused on the dominance of scientific "experts" and disregarded the "non-scientific rest of the world." But at the latest in "Science in a Free Society" (Feyerabend, 1980; original: 1978) and also in the posthumously published work "Conquest of Abundance. A Tale of Abstraction versus the Richness of Being" (1999b, p. 159; German translation: 2005), it becomes clear that he is concerned with more than that, that he opposes the narrow-mindedness and intolerance with which individuals and groups ignore the diversity of ways of life and worldviews. In "Conquest of Abundance," Feyerabend writes, for example, in essence, that today people can build worldviews on the basis of their individual decisions and thereby develop a commonality that was previously separated by the chauvinism of special groups (Feyerabend, 1999b, p. 159).[6]

However, the question arises as to what extent a modern concept of chauvinism is suitable for identifying those absurdities in science or everyday life against which Feyerabend takes a stand. As is often the case, Feyerabend refuses to precisely define the attributions and terms with which he characterizes people, views, or perspectives; whereby, in the Feyerabendian sense, one would have to ask what "precise" would mean—outside of any frame of reference.[7] The examples he provides to draw attention to the "destruction of diversity" rather describe forms, one could say, of narrow-mindedness *and* chauvinism. However, caution is warranted here as well: words like *narrow-minded* or *narrow-mindedness* he

[6] "It shows fear, indecision, a yearning for authority, and a disregard for the new opportunities that now exist: we can build worldviews on the basis of a personal choice and thus unite, for ourselves and for our friends, what was separated by the chauvinism of special groups" (Feyerabend, 1999b, p. 159). In the German edition, however, "Chauvinism" is somewhat inappropriately translated as "Untertanengeist" (Feyerabend, 2005, p. 173).

[7] That is, for Feyerabend, "precise" or "precision" are relative terms like any others, whose meaning can only be understood within the context of a particular (theoretical) frame of reference. Moreover, slavish adherence to the concepts and rules of a closed system should be avoided as much as possible (Feyerabend, 1987, p. vii). And: "Every rule, every law, even the most precise formulation dealing with events carefully prepared is bound to have exceptions…" (Feyerabend, 1970, p. 176).

hardly ever uses, and the combative term *chauvinism* only rarely in an explicit way.

To search for examples that, according to Feyerabend, indicate a narrow-minded or chauvinistic approach to diversity, and to do justice to the title of this book, for now a vague notion of narrow-mindedness and chauvinism will suffice.

At this point, in my view, it is not at all necessary to undertake a detailed etymological explanation and, for example, refer to the young Nicolas Chauvin from Napoleon Bonaparte's Grand Armée, whose name became synonymous with nationalist chauvinism. Nor is it necessary to recall at length Karl Marx, who used the term chauvinism to criticize the excessive nationalism with which people in European countries were prepared for mutual wars (Marx, 1965, p. 222; original: 1866; 1962, p. 558; original: April/May 1871). Hannah Arendt, for whom chauvinism represents the bridge between nationalism and imperialism (Arendt, 1945), should also be mentioned only briefly, as should current studies on chauvinism as an "aggressive national sentiment" (e.g., Decker & Brähler, 2020; Zick & Küpper, 2021).

One of the most apt definitions of "nationalist chauvinism" comes, incidentally, from Henri Dunant, the founder of the Red Cross, and Countess Hedwig Pötting, a friend and comrade-in-arms of Bertha von Suttner: "Chauvinism fosters hatred among peoples, nations, and races. One must not confuse this pernicious chauvinism with healthy patriotism. Patriotism is love, true devotion; chauvinism, on the other hand, is hatred—detestable fanaticism. Chauvinism glorifies brute force, it preaches injustice in the so-called 'right of occupation,' which civilized nations claim over those they call 'backward peoples'" (Dunant & Pötting, 1897, p. 311).

It is enough to recall that Paul Feyerabend repeatedly criticized the epistemic as well as political marginalization of indigenous peoples, the human and ecological costs of Westernization, the prevailing ideology of the "Western" conquerors, slavery, or wars, to get an idea of what he, Feyerabend, would describe as chauvinistic attempts to destroy diversity. In social-psychological terms, chauvinism could then be understood as a complex pattern of in-group favoritism and out-group discrimination, as exaggerated feelings and notions of the superiority of one's own group or community on the one hand, and the disparagement, discrimination, and contempt of other ("foreign") groups and communities on the other.

"Narrow-mindedness" is also a word with a broad semantic field. It was borrowed in the 18th century from the French *borné* and the verb *borner* ("to limit," "to restrict")—a derivative of the French *borne* ("boundary stone")—and means something like "close-minded"; as can be read in Friedrich Kluge's "Etymological Dictionary of the German Language" (1989, p. 98; original: 1881). Goethe used the word "borniert" and is said, for example, to have spoken of a "narrow-minded mass" (Eckermann, 1987, p. 665; original: 1848). It was apparently also one of the favorite words of Karl Marx and Friedrich Engels. In "The German Ideology," for example, they write of the "narrow-minded attitude of people toward nature" (Marx & Engels, 1969, p. 31; original: 1845/46) or of "the narrow-mindedness of the individuals themselves" (ibid., p. 67); in "Capital," the working method of Thomas R. Malthus is described as "narrow-minded" (Marx, 1977, p. 663; original: 1867), and so on.

Narrow-mindedness is less than chauvinism. Narrow-minded are those philosophers of science or ordinary people who, as Feyerabend reports, claim their own philosophy of science or worldview, their own ways of thinking, experiencing, and acting, exclusively as the only true and decisive instances for engaging with the world, and *ignore* the world of others and their perspectives (their myths, legends, or rituals). The narrow-minded think, feel, and act exclusively within their own filter bubble, to use the term introduced by Eli Pariser (Pariser, 2011), while the chauvinists *idealize* themselves and *devalue* and *despise* others.

So we are in good company in searching for narrow-minded and chauvinistic attempts to deny social, cultural, or scientific diversity. That will be the subject of the next part of this book.

References

Arendt, H. (1945). Imperialism, Nationalism, Chauvinism. *The Review of Politics, 7*(4), 441–463.

Bammé, A. (2006). Wissenschaft der Zukunft: Zukunft der Wissenschaft. Von der akademischen zur postakademischen Wissenschaft. In E. Buchinger & U. Felt (eds.), Technik- und Wissenschaftssoziologie in Österreich. *Österreichische Zeitschrift für Soziologie, Sonderheft 8,* pp. 277.

Boockmann, F., Bussotti, P., Di Liscia, D., & Oestmann, G. (2008). „Nicht das Kindt mit dem Badt außschütten". Zur Rolle einer Pseudowissenschaft im Zeitalter der wissenschaftlichen Revolution: Die Astrologie bei Johannes Kepler, Heinrich Rantzau und Galileo Galilei. *Bayerische Akademie der Wissenschaften, Akademie Aktuell*, 04, pp. 51–60. https://badw.de/fileadmin/pub/akademieAktuell/2008/27/19_DiLicia.pdf; Accessed: 4. Mar. 2023.

Brückler, F.M. (2017). *Geschichte der Mathematik kompakt.* Springer Spektrum.

Camus, A. (1950). Der Künstler und die Freiheit. *Der Monat. Internationale Zeitschrift für Politik und geistiges Leben*, *3*(017), pp. 522–526. https://www.ceeol.com/search/article-detail?id=93466; Accessed: May 15, 2023.

Camus, A. (1951). Nietzsche und der Nihilismus. *Der Monat. Zeitschrift für Politik und geistiges Leben, 4*(039), pp. 227–236; https://www.ceeol.com/search/viewpdf?id=68628 Accessed: May 15, 2023.

Camus, A. (1997). *Original: 1942).* Der Mythos von Sisyphos. Ein Versuch über das Absurde: Rowohlt.

Camus, A. (2011; Original: 1951). *Der Mensch in der Revolte*. Rowohlt.

Decker, O., & Brähler, E. (2020*). Autoritäre Dynamiken. Alte Ressentiments—neue Radikalität.* Psychosozial-Verlag.

Dunant, H., & Pötting, G. H. (1897). Kleines Arsenal gegen den Militarismus. *Die Waffen nieder!*, *6*(8/9), pp. 310–314.

Eckermann, J. P. (1987). *Original: 1848).* Gespräche mit Goethe: Aufbau-Verlag.

Feyerabend, P. K. (1963). How to be a Good Empiricist: A Plea for Tolerance in Matters Epistemological. In B. H. Baumrin (ed.), *Philosophy of Science: The Delaware Seminar* (vol. 2, pp. 3–39). Interscience.

Feyerabend, P. K. (1968). On a Recent Critique of Complementary. *Philosophy of Science, 35*(4), 309–331.

Feyerabend, P. K. (1970). Philosophy of Science: A Subject with a Great Past. In H. Stuewer (ed.). *Historical and Philosophical Perspectives of Science (Minnesota Studies in the Philosophy of Science), 5*, 172–183.

Feyerabend, P. K. (1978). *Der wissenschaftstheoretische Realismus und die Autorität der Wissenschaften*. Vieweg & Sohn.

Feyerabend, P. K. (1980; Original: 1978). *Erkenntnis für freie Menschen*. Suhrkamp.

Feyerabend, P. K. (1981). *Realism, Rationalism and Scientific Method: Philosophical Papers* (vol. 1). Cambridge University Press.

Feyerabend, P. K. (1984). *Wissenschaft als Kunst.* Suhrkamp.

Feyerabend, P. K. (1986; Original: 1975). *Wider den Methodenzwang.* Suhrkamp.

Feyerabend, P. K. (1987). *Farewell to Reason.* Verso.

Feyerabend, P. K. (1992; Original: 1989). *Über Erkenntnis. Zwei Dialoge.* Campus.

Feyerabend, P. K. (1995a). *Zeitverschwendung*. Suhrkamp.

Feyerabend, P. K. (1995b). *Briefe an einen Freund, herausgegeben von Hans Peter Duerr.* Suhrkamp.

Feyerabend, P. K. (1999a). *Knowledge, Science and Relativism, herausgegeben von John Preston*. Cambridge University Press.

Feyerabend, P. K. (1999b). *Conquest of Abundance. A Tale of Abstraction versus the Richness of Being*, herausgegeben von Bert Terpstra. The University of Chicago Press.

Feyerabend, P. K. (2005). *Die Vernichtung der Vielfalt. Ein Bericht*, herausgegeben von Peter Engelmann. Passagen.

Feyerabend, P. K. (2009). *Naturphilosophie,* herausgegeben von H. Heit und E. Oberheim. Suhrkamp.

Guilford, J. P. (1967). *The Nature of Human Intelligence*. McGraw-Hill.

Hagner, M. (2017). Wider den Populismus. *Paul Feyerabends dadaistische Erkenntnistheorie. Zeithistorische Forschungen, 14*(2), 369–375.

Heit, H. (2016). Wissenschaftskritik in der Genealogie der Moral. Vom asketischen Ideal zur Erkenntnis für freie Menschen. In H. Heit & S. Thorgeirsdottir (eds.), *Nietzsche als Kritiker und Denker der Transformation* (pp. 252–274). De Gruyter.

Hoyningen-Huene, P., & Oberheim, E. (2006). Neues zu Feyerabend. In F. Stadler & K. R. Fischer (eds.), *Paul Feyerabend. Ein Philosoph aus Wien* (pp. 13–33). Springer.

Kidd, I. J. (2011). Objectivity, abstraction, and the individual: The influence of Søren Kierkegaard on Paul Feyerabend. *Studies in History and Philosophy of Science Part A, 42*(1), 125–134.

Kluge, F. (1989). *Original: 1881).* Etymologisches Wörterbuch der deutschen Sprache: De Gruyter.

Marx, K. (1962; Original: 1871). Erster Entwurf zum „Bürgerkrieg in Frankreich". In *Karl Marx & Friedrich Engels, Werke, vol. 17.* Dietz.

Marx, K. (1965; Original: 1866). Marx an Engels in Manchester, 7. Juni 1866. In *Karl Marx & Friedrich Engels, Werke, vol. 31.* Dietz.

Marx, K., & Engels, F. (1969; Original: 1845/46). Die deutsche Ideologie. In *Karl Marx & Friedrich Engels, Werke, vol. 3.* Dietz.

Marx, K. (1977; Original: 1867). Das Kapital. In *Karl Marx & Friedrich Engels, Werke, vol. 23.* Dietz.

Neto, J. R. M. (1991). Feyerabend's Scepticism. *Studies in History and Philosophy of Science Part A, 22*(4), 543–555.

Pariser, E. (2011). *The filter bubble: What the Internet is hiding from you.* Penguin.

Piper, H. J. (2002). Zarathustra- Sisyphos. Zur Nietzsche-Rezeption Albert Camus '. Nietzscheforschung, vol. 9 (pp. 247–261), herausgegeben von Volker Gerhardt and Renate Reschke, Berlin: Akademie Verlag; https://doi.org/10.1524/9783050047294.247.

Preston, J. (2002). "Paul Feyerabend". The Stanford Encyclopedia of Philosophy (Summer 2002 Edition); http://plato.stanford.edu/archives/sum2002/entries/feyerabend. Accessed: Jan. 14, 2023.

Reif, A. K. (1999). „Die Welt bietet nicht Wahrheiten, sondern Liebesmöglichkeiten". Zur Bedeutung der Liebe im Werk von Albert Camus. Dissertation. Universität-Gesamthochschule Wuppertal. http://elpub.bib.uni-wuppertal.de/edocs/dokumente/fb02/diss1999/reif/d029902.pdf. Accessed: March 15, 2023.

Rohr, C. (2013). Macht der Sterne, Allmacht Gottes oder Laune der Natur? Astrologische Expertendiskurse über Krisen und Naturrisiken im späten Mittelalter und am Beginn der Neuzeit. In C. Meyer, K. Patzel-Mattern & G. Jasper Schenk (Hrsg.), *Krisengeschichte(n)—„Krise" als Leitbegriff und Erzählmuster in kulturwisssenschaftlicher Perspektive* (pp. 361–385). Steiner.

Saltzwedel, J. (1999). Märchen des Geistes. Der Spiegel, 14. Februar 1999. https://www.spiegel.de/kultur/maerchen-des-geistes-a-c01b7bbb-0002-0001-0000-000009032710. Accessed: March 3, 2023.

Suntrup, J. C. (2009). Eine intellektuelle Ethik der Bescheidenheit und der Solidarität. Albert Camus' Denken der Revolte. *Zeitschrift für Politik, Vol. 56*, 2, pp. 179–196.

Zick, W., & Küpper, B. (eds.) (2021). *Die geforderte Mitte. Rechtsextreme und demokratiegefährdende Einstellungen in Deutschland 2020/21.* J.H.W. Dietz Nachf.

Zoglauer, T. (2021). *Konstruierte Wahrheiten: Wahrheit und Wissen im postfaktischen Zeitalter.* Springer Vieweg.

Part V
From the Logbook of the Absurd

"Occasionally, one is almost inclined to say: as many sci0 The philosopher of science died in Vienna" (taz 1994). Well, as is well known, he did not die in Vienna, but in a Swiss hospital, but he left us too soon. A year before his death, he said—probably not entirely seriously—in an interview with the Swiss "WOZ. Die Wochenzeitung" that nothing in this world was more certain than death, "[...] and not even that is certain, because who knows whether I won't come back as a ghost and give another interview" (Feyerabend, 1993; quoted in Rapp Wagner, 1997).

What would he say in the face of the absurdities of our present time? It is hard to say. One could, however, ask whether the social and scientific pluralism demanded by Paul Feyerabend is an appropriate means or a convincing approach to counter narrow-mindedness and chauvinism in dealing with scientific and social problems and conflicts. This will be exemplified in this part of the book. Paul Feyerabend's ideas serve here as (not always valid) blueprints.

The archives documenting the absurdities of contemporary times are amply filled. Consider, for example, to begin with the seemingly trivial, gendering (in the German language) —with asterisk, with capital I, through neutralization. How do I gender correctly? Is this the burning question of our time? Hardly. What about psychology? Is it once again in crisis, or is it the "modern opium of the people" (as the subtitle of a provocative book by Albert Krölls, 2016, puts it)? Are the psychological findings of recent decades—turning to more pressing questions—universally valid, or merely products of the "West"? Is the "West" not finished anyway, because it is Eurocentric and imperialistic, or can its Enlightenment attitudes still be salvaged, as some representatives of postcolonial studies are eagerly debating? What are the "corona deniers" actually up to? Are they still spreading their conspiracy narratives, like the climate deniers and their female and male companions? Although, to speak and write of climate deniers is imprecise and inappropriate. These so-called deniers do not doubt that there is a climate. But "climate change deniers" does not seem entirely accurate either. Strictly speaking, they should be called "climate change cause diagnosis consensus skeptics," as Jens Soentgen and Helena Bilandzic (2014, p. 42) somewhat mockingly suggest. Gendering becomes difficult in that case, however. Was the flood disaster that occurred in 2021 in Rhineland-Palatinate and North Rhine-Westphalia, in which 133 people died in the Ahr Valley, artificially caused? At least, that is what a Bundestag candidate from the party "Die Basis" claimed in 2021 (see also: BR24, 2021). Can "artificial intelligence" be human? Are there AfD members who are democrats? Are attitudes and behaviors toward Jews the litmus test of any form of humanism? And: In light of Russia's war of aggression against Ukraine, is it still possible to be a pacifist? And so on, and even more questions come to mind. For: "Great is the confusion under heaven. Strange things are happening in these times" (Eco, 2012, p. 213). Not all strange "things" from the archives of absurdities can be observed, not all questions can be asked, and only a few answers can be formulated—on the one hand due to lack of expertise, on the other due to the space available in this book. I have, however, already attempted to answer some questions (Frindte, 2022, especially Part V). Now I shall limit myself.

References

BR24. (2021). Faktenfuchs: The floods were not artificially triggered. https://www.br.de/nachrichten/wissen/faktenfuchs-die-hochwasser-wurden-nicht-kuenstlich-ausgeloest,SdzgGfx. Accessed: June 28, 2023.

Eco, U. (2012). *Advancing Sideways*. Deutscher Taschenbuch Verlag.

Feyerabend, P. K. (1980). *Knowledge for Free People*. Suhrkamp.

Feyerabend, P. K. (1993). Principles are tombstones. Interview in "WOZ. Wochenzeitung" from January 29, 1993; cited in Renata Rapp Wagner (1997, no page reference), Postmodern Thinking and Pedagogy. A Critical Analysis from a Philosophical-Anthropological Perspective. https://naturrecht.ch/renata-rapp-wagner-postmodernes-denken-und-paedagogik-eine-kritische-analyse-aus-philosophisch-anthropologischer-perspektive/. Accessed June 15, 2023.

Frindte, W. (2022). *Quo Vadis, Humanism*? Springer.

Krölls, A. (2016). *Critique of Psychology. The Modern Opium of the People*. VSA.

Soentgen, J., & Bilandzic, H. (2014). The Structure of Climate-Skeptical Arguments. *Conspiracy Theory as Critique of Science. GAIA-Ecological Perspectives for Science and Society*, *23*(1), 40–47.

taz. (1994). https://taz.de/Paul-Feyerabend-als-Geist-zurueck-Ja-Bitte-Sofort/!1576348/. Accessed June 5, 2023.

"Psychology—God Have Mercy, Are You Still Sticking With It?" 14

Thus the devil to Adrian Leverkühn: "Psychology—God have mercy, are you still sticking with that? That's just bad, bourgeois nineteenth century! The era is thoroughly sick of it, soon it will be a red rag to it, and anyone who disturbs life with psychology will simply get a blow to the head" (Mann, 1954, p. 339; original: 1947). Well, it didn't turn out quite that bad. Nevertheless, some developments are indeed cause for concern. That's why this chapter turned out a bit longer.

14.1 Psychology in Crisis Mode?

"Intellectuals do not solve crises, they create them" (Eco, 2012, p. 63).

At times, one might think that the history of modern psychology is a history of crises and controversies (Galliker, 2016). At the turn from the 19th to the 20th century, psychologists argued in the context of a "crisis of subject matter" about the object of their discipline and the methods for its investigation. For example, Wilhelm Dilthey (1833–1911) and Herrmann Ebbinghaus (1850–1909) were deeply divided between 1894 and 1896 over whether psychology should be pursued as a humanities-oriented interpretive psychology or as a natural science-oriented experimental psychology (Ebbinghaus, 1896; Dilthey, 1894). A similar controversy existed between Wilhelm Wundt and the protagonists of the Würzburg school of thought psychology (e.g., Bühler, 1908; Wundt, 1907). Superficially, this controversy may seem to have been merely a dispute over methodological questions, but it also touches on the fundamental issues of a future psychology: Are the subjects in psychological experiments merely

W. Frindte, *Against Narrow-Mindedness and Chauvinism—Journeying Through Absurd Times with Paul K. Feyerabend*, https://doi.org/10.1007/978-3-658-50113-6_14

"objects" of research, exposed to certain stimuli to test their reactions? Or are the "participants" subjects of their own thinking and actions, capable of consciously reflecting on their experimental performances (memory tasks)? In 1927, Karl Bühler explicitly wrote about the crisis in psychology (Bühler, 1927), arguing that psychological science was facing a "crisis of structure," which, while revealing the theoretical and methodological inconsistencies among the discipline's representatives, would ultimately lead to a unified science. As is well known, this did not happen.

After Hitler's rise to power and the ascent of the NSDAP to the ruling party, the darkest chapter in the history of German-speaking psychology began. To call this "chapter" a crisis would be euphemistic. On April 7, 1933, the National Socialists passed the "Law for the Restoration of the Professional Civil Service," as a result of which "non-Aryan" and politically undesirable civil servants were forced into retirement. This affected all Jewish professors. By the end of 1933, 313 full and 109 associate professors, about 400 honorary professors and private lecturers, as well as more than 500 staff members at scientific institutes, museums, and libraries had been dismissed from German universities (Jaeger, 1993, p. 221). One third of all full professors and numerous senior teaching staff involved in psychological teaching and research lost their positions due to the persecution of Jews. This effectively silenced progressive German psychology. Of the 308 members of the *German Society for Psychology* living in the German-speaking world in 1932, a total of 45 emigrated from 1933 onward (see Ash, 1985, p. 74). Among them were most of the leading psychologists of the time: Curt Bondy (Göttingen), Charlotte and Karl Bühler (Vienna), Jonas Cohn (Freiburg), Adhémar Gelb (Halle), Erich von Hornbostel (Berlin), David Katz (Rostock), Wolfgang Köhler (Berlin), Kurt Lewin (Berlin), Wilhelm Peters (Jena), Otto Selz (Mannheim, later arrested by the Nazis in his Dutch exile and murdered in Auschwitz), William Stern (Hamburg), Max Wertheimer (Frankfurt), Traugott E. K. Oesterreich (Tübingen), Heinz Werner (Hamburg), Karl Duncker, Helene Frank, Liselotte Frankl, Else Frenkel-Brunswik, Marie Jahoda, Paul Felix Lazarsfeld, the psychoanalysts Edith Buxbaum, Frieda Reichmann-Fromm, and the psychoanalysts Alfred Adler, Siegfried Bernfeld, Otto Fenichel, Sigmund Freud, Ernst Fromm, Wilhelm Reich, and many others (see also: Wolfradt et al., 2017).

At the beginning of the 1950s, some psychologists in the former Federal Republic once again found themselves confronted with a crisis. This time, it was about the influence of American psychology in teaching and research, about the "Americanization" of German psychology, as Alexander Métraux later called it (Métraux, 1985, p. 246).

The crisis discussions did not end. In the 1970s, there was talk of a "crisis of social psychology" (Mertens & Fuchs, 1978), one in psychodiagnostics (Pulver, Lang & Schmid, 1978), a "crisis of methodology" (Maschewsky, 1977). At the beginning of the 1990s, psychology once again seemed to be in crisis mode. Heiner Leggewie called for a "renewal of psychology" (Legewie, 1991), arguing that "nomological" psychology—that is, psychology focused on the testing of law-like statements—was irrelevant to the concerns of practicing psychologists. Theo Herrmann countered, pointing out that research problems should not be conflated with crises in the training and application of psychology. He did offer a "pluralism of theories" (see also Herrmann, 1979), but limited the pluralistic framework to nomological methodology and analytic philosophy of science (Herrmann, 1991). It was obvious that agreement was neither possible nor desired. The dispute over "renewal" (Legewie) versus "improvement" (Herrmann) resembled a zero-sum game: each claimed to possess the epistemological positions from which psychology could be practiced exclusively. In the end, the reformers distanced themselves from "nomological psychology" and its institution, the *German Society for Psychology,* and founded a new association, the *New Society for Psychology* (see also Volmerg, 1992). Paul Feyerabend would call it *scientific chauvinism.* I call it *narrow-mindedness,* because in all the debates about renewal or improvement, one truly grave crisis was completely overlooked: the crisis of GDR psychology. One can debate the nature of this GDR psychology and the extent of its international reputation. It cannot be reduced to its instrumentalization by the SED dictatorship and the Ministry for State Security (see also Bredenkamp, 1991; Busse, 1993; Frindte, 2022; Maercker & Gieseke, 2021).

Nevertheless, not much remains of the former GDR psychology, especially not of its social psychological approach. Certainly, the influential works of Friedhart Klix on cognitive psychology (e.g., 1971, 1984) or of Winfried Hacker on occupational psychology (e.g., 1973, 2021) still have great potential for further development today. The situation is quite different in social psychology. It belonged to that part of GDR psychology most closely intertwined with the party and state regime. Numerous official and unofficial employees of the State Security had studied social psychology in Jena (sometimes together with their later victims), received training and further education at the Stasi college in Potsdam-Golm, or even worked themselves as operative psychologists[1] (see in detail: Lenski, 2021). But that alone cannot be the reason why GDR social psychology has either fallen into oblivion, is being ignored in remarkable unanimity, or is only perceived in terms of its entanglement with the SED state.

Perhaps such judgments as those made in the 1990s, for example by William Woodward and Steven Clark (1996), led to the narrow-minded rejection of GDR social psychology: The "East German" psychology, if one looks at the subject index in the social psychology textbook by Hiebsch and Vorwerg (1979), remained profoundly monocultural, nationalistic, and ethnocentric (Woodward & Clark, 1996, p. 240); although methods of multivariate statistics and mathematical modeling were used, no other methodological techniques were employed (ibid., p. 246); experimental research lagged behind international developments (ibid., p. 247); finally, East German psychologists had no choice but to either adopt Western research methods or retire.

As is well known, a textbook does not necessarily reflect the latest developments in a scientific discipline. Setting aside politically ideological gestures of compliance, for example, "Wissenschaftspsychologie" (Hiebsch, 1977) or the collaborative work "Interpersonelle Wahrnehmung und Urteilsbildung" (Hiebsch et al., 1986) were certainly up to date. Well, these are the complaints of an old white man.

While innovators in West Germany emancipated themselves from the reformers, psychologists in the academic institutions of East Germany were swept away by the "waves of dissolution and renewal" in the early 1990s. These "waves," in which the institutional structures of psychology and their personnel were called into question or evaluated for their professional competence, oscillated between necessary and appropriate intentions of assessment and hasty, sometimes prejudiced, interests in devaluation. Mitchell G. Ash has attempted, with some historical distance, to trace the radical changes in "East German" psychology after 1989. "West Germanification," as one of his conclusions, that is, the restructuring of East German universities according to the West German model, was of a destructive force (Ash, 2023, p. 12).

[1] The "operative psychology" of the Ministry for State Security (MfS) of the GDR was, on the one hand, the name for the psychology curriculum at the MfS college in Potsdam. On the other hand, this term refers to the actual application of psychological tactics in the everyday practice of the Stasi, as can be documented by many victim files.

Approximately 60% of scientists at GDR universities and more than 80% of scientists in industrial research lost their jobs by 1992. A smaller portion was dismissed because they had worked as informal collaborators for the state security service or were classified as system-loyal. The much larger portion was—unpleasantly put—terminated due to a lack of professional competence. The criteria for these professional evaluation processes were set by representatives of West German academic disciplines, who generally also undertook the scientific assessment of East German scientists. The suspicion that the dissolution for professional reasons also had the effect of creating academic spaces for the second and third tier of West German scientists is not unfounded.

Setting aside the aforementioned crisis of GDR psychology, which is far from being comprehensively researched, and considering the psychology crises mentioned as representative, the conclusion is obvious: The controversies over the past decades regarding "natural science versus humanities" psychology, "quantitative" versus "qualitative" research, "object-" versus "subject psychology," or mathematization versus understanding of meaning, etc., all belong to the attempts to deconstruct the *myth* of "objective knowledge" even in psychological science and, as Feyerabend wrote in the passage already mentioned, to help the "subjective" side of knowledge get back on its feet (Feyerabend, 2005, p. 158; original: 1989).

It would be too much speculative optimism to assume that Paul Feyerabend had similar discussions in psychology in mind with his "subjective elements." Nevertheless, let us recall the debates and controversies that circulated from the 1960s into the 1980s. One might call it the "rediscovery of the subject" that took place (Gummersbach, 1985, pp. 324). At first, this may sound like a tautology, assuming that psychology is a science of the subject. However, in the tradition of positivism and behaviorism, the subjectivity of humans in psychological research had largely been reduced to their responsiveness in experimental settings and controlled through sophisticated techniques. In the 1960s, discussions began about the pros and cons of experimental designs and their closeness to reality (e.g., Timaeus, 1974), the hardly controllable influence of researchers on the subjects being studied (the so-called experimenter effects; Rosenthal, 1966) was problematized, or the realization grew that subjects (as "participants") indeed have their own and often very different motivations ("participant motivations") for taking part in psychological research (Orne, 1962). There was discussion—later in German-language psychology—about the so-called artifacts (Kriz, 1981), that is, the artificial products, the "erroneous research results," and ways were sought to get a handle on the unruly subject in everyday research (e.g., Bungard, 1984; Greenwood, 1983).

Against the backdrop of a potential multitude of confounding factors, proposals soon multiplied on how to deal with such confounding factors in (social) psychological experiments. Some researchers emphasized the general limits of the experimental method (e.g., Bungard, 1984). Others made suggestions to make it more difficult for "participants" to discern the "true" intentions of the experimenter and thus eliminate "undesired participant expectations" (e.g., through the use of non-reactive procedures[2]; Webb et al., 1975). Still others advocated for making the interactions between experimenter and "participants" more open and equal, forgoing deception, and codifying ethical principles in dealing with "participants" in order to encourage them to disclose more and better information about themselves (e.g., Kelman, 1967, Schuler, 1980). There were also calls for a complete abandonment of conventional laboratory experiments in favor of role-playing and field studies (e.g., Greenwood, 1983; Mertens, 1980). In any case, the discussions among artifact researchers revealed that apparently the entire (social) psychological research process is potentially artifact-laden, that artifacts tend to occur not only in psychological experiments, and that there seem to be hardly any means or ways to avoid artifacts.

By the mid-1980s, discussions about the susceptibility of psychological research to artifacts had subsided. Some gave up in resignation, others turned to non-experimental, qualitative research (see, e.g., Mayring, 1990). In some branches of psychological science, the realization emerged that more precise research methods were not needed, but rather that the debate about the "object" of psychology was on the agenda. Humanity and subjectivity moved to the center of these debates (e.g., Gergen, 1989; Groeben & Scheele, 1977; Holzkamp, 1983).

Meanwhile, the herd of the mainstream continued on its "via regia."

Some psychologists declared the crisis in psychology to be over, the methodological problems as manageable, and the supposed ethical issues of psychological experimentation as recognized. For example, they saw the study of "social cognitions" as a new paradigm in social psychology. Behind this paradigm lies the conviction that behavior is not the appropriate level of analysis for (social) psychology, but that social behavior can only be satisfactorily explained if the mental processes governing behavior are sufficiently understood (e.g., Strack, 1988, p. 73). Others limited themselves to the development and testing of small

[2] Non-reactive methods of data collection are procedures in which the "studied" individuals are not aware that they are participating in a study.

or medium-sized theories. At the beginning of the new millennium, Arie Kruglanski lamented not only a general theoretical reticence ("theory shyness") among (social) psychologists, but also criticized the efforts initiated by Robert Merton (1957) to construct and empirically test primarily "middle-range" theories (Kruglanski, 2001). Another segment of the mainstream is dedicated to refining methodological approaches. Data collection methods have multiplied. For instance, mixed methods—the combination of qualitative and quantitative data collection procedures—have gained importance in recent years (Schreier & Odağ, 2020; in practical projects, e.g., Frindte & Haußecker, 2010), even though psychologists working qualitatively or hermeneutically still have to defend themselves against the narrow-mindedness and marginalization by their "natural science"-oriented colleagues (e.g., Schreier & Odağ, ibid., p. 162). The experiment is still regarded by some psychologists as the "royal road of causal analysis" (Bohner, 2016, p. 58). However, panel studies—longitudinal studies with the same individuals over several points in time, which can be analyzed using sophisticated multivariate techniques—are increasingly developing into challenging competing enterprises (Pforr & Schröder, 2015; in practical projects, e.g., Eyssel et al., 2015).

And finally, there is now the group of psychologists who—much like Feyerabend's psychology teacher Hubert Rohracher years ago—base their psychological research on the neurosciences in order to find the mental in neural processes, using, among other things, electroencephalography, magnetic resonance imaging, functional near-infrared spectroscopy, or eye-tracking. Mark Galliker therefore anxiously asks whether psychology can still exist alongside the overpowering neurosciences (Galliker, 2016, pp. 15 ff.). Others, however, are of the opinion that psychology and neuroscience are not competing disciplines, but should regard each other as auxiliary sciences (e.g., Bak, 2021, p. 8).

The mind-body problem, as Feyerabend foresaw, is still very much alive, and the "subjective" side of knowledge simply cannot be eradicated.

14.2 Is That Narrow-mindedness?

> "Some people knock around on the wall with a hammer and believe they hit the nail on the head every time." (Goethe, Maxims and Reflections, 1972, p. 493; original: 1820).

Actually, psychology as a science is doing quite well, if you look at the exponential growth of scientific publications in recent decades.[3] So things are moving forward, even if quantity says nothing about the quality of the respective publications. But that seems to be the problem—or at least one of many. According to Wolfgang Maiers (2021), after medicine and the life sciences, psychology too has been experiencing a replication crisis since 2010.

Replication refers to the repetition of studies in order to confirm, refute, or extend their results. The aim of repetitions is thus to demonstrate the validity and stability of the originally produced findings, and ultimately, to provide credible, reliable, and useful scientific knowledge.

By the way, ChatGPT knows this too. On June 19, 2023, the chatbot gave me the following answer to my question about whether it could tell me something about the replication crisis: "The replication crisis refers to the difficulties many scientists have in repeating experiments and studies. There is growing concern that many scientific results are not reproducible and that this can lead to a distortion of scientific knowledge. There are various reasons for the replication crisis, including a lack of transparency in reporting research results, inadequate statistical methods, and an excessive emphasis on significant results. The scientific community is working to find solutions to these problems in order to improve the quality and reliability of scientific knowledge." Not bad at all.

The so-called replication crisis began with the realization that psychologists are, after all, only human—intelligent, lovable, but also selfish, deceitful, and much more. In 2012, a team led by the prominent psychologist and behavioral economist Dan Ariely published a study on honesty (Shu et al., 2012). The reported finding was as simple as it was convincing: when people, for example, have to sign tax returns, insurance policies, or other official documents to confirm that they have answered honestly, the signature should be required at the beginning rather than at the end of the documents. This, it was claimed, increases honesty. One can imagine that the study was not only frequently cited but also praised and welcomed by companies, tax authorities, insurance companies, etc. There was

[3] Whether the same can be said of its application will not be addressed here.

only one problem, which emerged in 2020: not only could the study not be replicated; the underlying data were fabricated, and Dan Ariely may have been the main fraudster (see also Diekmann, 2021).

The absurd thing is that at the beginning of the 2010s, others were also cheating and falsifying data. For example, Diederik Stapel, a prominent, much-cited, and award-winning Dutch social psychologist. Of the works he published in reputable journals, at least 55 are said to have been based on falsified data. For years, Stapel had attracted attention with supposed research findings such as: meat-eaters are less sociable and more selfish than vegetarians; women should think twice about taking their husband's name after marriage, because then they are considered less intelligent, whereas if they keep their own name, they are seen as more ambitious and also less caring (see also Konitzer, 2013, p. 66). Stapel's own doctoral students uncovered the fraud (Rost & Bienfeld, 2019).

Well, cheating in science is nothing new. Perhaps you know the example of Cyril Burt, one of the pioneers of applied psychology in England. He was a professor of psychology at University College London and was the first psychologist to be knighted; in 1971, the American Psychological Association awarded him the Thorndike Prize (for the first time, this high honor went to a foreigner). The only problem with Burt: he was a fraud. He fabricated data to support his theories. He used his masterful command of statistics to fool both his admirers and his critics. In his time, Burt published IQ data from identical twins raised apart. It was the largest collection of its kind in the world. Since his twins had the same genetic makeup but came from different environments, they could serve as ideal subjects for studying the interaction of heredity and environment. Based on his findings, Burt concluded that intelligence was predominantly determined by genetics. When he died in 1971 at the age of 88, the fraud came to light. The whole affair was uncovered by a psychologist from Princeton, Leon Kamin. In 1972, a student urged him to look at Burt's papers (Kamin, 1979; original: 1974). This and a later review of Burt's diaries revealed something very damning: Burt had not conducted the research he had claimed and published for decades. Twin research had its scandal. The falsified data did not at all prove the overwhelming influence of genetic factors on the development of intelligence, but showed something many had long suspected: scientists, too, are only human. As Paul Feyerabend once said: "You simply can't rely on scientists ..." (Feyerabend, 1980, p. 188).

Incidentally, Paul Feyerabend was quite familiar with the fabrications of Sir Cyril Burt, but maintained that Burt's deceptions, compared to certain psychological schools (e.g., behaviorism), which place little value on personal experience and disenfranchise individuals, were childish and harmless (Feyerabend & Thomas, 1986, p. 216).

The now-declared replication crisis, however, strikes at the very heart of the matter.

In a large and systematic investigation by the Open Science Collaboration (2015), 100 experimental and correlational studies from three influential psychology journals (*Psychological Science, Journal of Personality and Social Psychology*, and *Journal of Experimental Psychology: Learning, Memory, and Cognition*) were analyzed. The results showed, among other things: While significant effects were demonstrated in 97% of the original studies, only 36% of the replication studies were actually able to confirm these significant effects. Other systematic analyses of international studies indicate a replication rate between 54 and 62% (see also Brachem et al., 2022, p. 2).

In addition, studies that found null effects—i.e., where the predicted positive relationships between the psychological variables under investigation could not be demonstrated—have little chance of being published in reputable journals. Not only would Karl Popper be turning in his grave, but those who still cling to naive falsificationism ought to be ashamed.

The Psychology Review Board of the German Research Foundation analyzed the situation and estimated that approximately 43% of published psychological studies report "true positive" results. Most, that is, 57%, of the reported significant results are thus false positives (Ulrich et al., 2016, p. 165). False positive findings can arise because researchers do not adhere to or deliberately violate methodological standards—for example, when they investigate the relationship between several "independent" and "dependent" variables but report only the significant relationships, or when they do not complete the data analysis but stop as soon as the desired significant relationships are found.

And then, as mentioned, there are also data fabrications and data manipulations, which account for about two percent of all publications in psychology (Fanelli, 2009). In neuroscience and medicine, however, the rate of "fake publications"—publications with falsified data—is likely even higher. Researchers at the University of Magdeburg analyzed approximately 15,000 publications from these scientific fields. According to the authors, between 2010 and 2020, the rate of "red-flagged fake publications" rose from 16% to 28%. A particularly high rate is evident in publications from Russia, Turkey, China, and Egypt (Sabel et al., 2023).

If one assumes that psychologists now have at their disposal not only Logical Empiricism or Critical Rationalism, but a colorful bouquet of foundational assumptions in the philosophy of science (from Thomas Kuhn to Paul Feyerabend to varieties of constructivism, e.g., Herzog, 2012; Walach, 2020), then one must ask: Cui bono? Are researchers even interested in philosophy of science, or do they, in *blinkered* arrogance, simply choose to ignore such foundations? On the one hand, Paul Feyerabend would be pleased about this; on the other hand, even he would have to ponder in light of the evidence that many fundamental psychological effects have not yet been replicated, and that falsifying or manipulating data apparently also belongs to the practice of scientific knowledge production.

In this case, the whiff of narrow-mindedness does not surround the philosophers of science, but rather the manufacturers of knowledge and their sponsors. They, or at least a significant portion of them, are guided not by the noble pursuit of knowledge, but by the desire for recognition, career advancement, and the need for validation.

The research funders, universities, foundations, journal publishers, and other institutions that provide research funding and publication opportunities expect "new" findings and less often the confirmation or refutation of what is already known. Even the second-order observers, such as the reviewers of manuscripts submitted for publication, may not be free from narrow-mindedness. They too are people with interests, convictions, and subtle biases. They often tend to favor scientific findings that align with their own views and reject work that does not correspond to their own state of knowledge or that of their scientific community. In psychology, this tendency is also known as *confirmation bias*.

> "Confirmation bias (also 'confirmation error') describes the tendency of individuals to seek, interpret, or recall information that confirms their opinion on a particular topic. This phenomenon was originally studied in the context of hypothesis testing: studies have shown that people generally try to confirm an existing hypothesis, but not to falsify it" (Peter & Brosius, 2013, p. 467).

Overconfidence, the tendency to overemphasize the weaknesses of a manuscript compared to its strengths, or the influence of (negative) prior information about the author can also cloud the reviewer's judgment (for a detailed discussion, see King et al., 2018).

14.3 A Hint of Chauvinism

> "An element of confusion is that it is often not possible to distinguish between identification with one's own roots, understanding those who have different roots, and judging what is good and what is bad" (Eco, 2012, p. 202).

There must have been times, and not so long ago, when people—whoever they were—believed that the psychology we know and practice is based on experiments with U.S. college students. Such an opinion, if it was ever actually held, can, provided one has a certain degree of knowledge of the history of psychology, safely be described as oblivious to history. On the one hand.

On the other hand, it is probably not entirely wrong to assume that the majority of psychological studies worldwide are based on samples drawn from Western, educated, industrialized, rich, and democratic societies: *Western, Educated, Industrialized, Rich* and *Democratic*, in short: WEIRD. An analysis of psychological journals from 2003 to 2007 found that 96% of the participants in the reported studies came from Western industrialized countries, including Israel and Australia. That is, 96% of the samples represented 12% of the world's population. In 67% of U.S. studies and in 80% of other studies, the samples consisted of psychology students (Arnett, 2008; cited in Henrich et al., 2010, p. 63).[4] This would not be quite so bad if some of the "weird" researchers did not also claim to generalize their findings about "weird" people to people in general. Perhaps these are exceptions. But Joseph Henrich, Steven Heine, and Ara Norenzayan (2010) have examined a whole series of such claims in great detail and conclude that psychology, if this continues to be common research practice, is facing an alarming situation. The samples of Western students are flooding the psychological databases; yet these samples represent the least suitable population for expanding our knowledge of Homo sapiens (Henrich et al., 2010, p. 83).

One seems to sense a hint of chauvinism when looking at the attempts to generalize psychological findings about the behavior of students in Western laboratories or other methodological settings to the "world as such" or to "humanity here and now." Not to forget Paul Feyerabend (see also *Chap. 11*): "Since people were discovered who did not belong to the Western cultural and civilizational sphere,

[4] To what extent these figures are valid is difficult to assess. I would prefer to set the percentages lower. My own database research for the year 2022 showed that in a maximum of 40% of all German-language experimental studies, the participants came from the student milieu.

it was almost considered a moral duty to bring them the truth—and by that was meant the prevailing ideology of the conquerors" (Feyerabend, 1992, pp. 52).

Admittedly, perhaps this is a bit too harsh in view of the efforts to simply overstate one's own research achievements a little and not to specify how representative one's own studies actually are. Let us take a closer look at some psychological effects as examples with regard to their generalizability:

Let us begin with the Müller-Lyer figure, briefly mentioned in *Chap. 12*, which can produce an optical illusion that might also have impressed Paul Feyerabend. It was discovered in 1889 and for a long time was considered a universal, phylogenetically preformed perceptual phenomenon. Until, in the 1960s, a team led by anthropologist Marshall H. Segall was able to show in cross-cultural comparative studies, including in South Africa, that this is not the case (Segall et al., 1966). The Müller-Lyer illusion, according to the results, apparently only occurs when the perceiving individuals have been socialized in cultural contexts in which "the everyday environment is characterized by straight lines, edges, and angles, so that the inward or outward pointing ends of the figure are perceived as the vanishing lines of a three-dimensional geometric object" (Breyer, 2017, p. 142).

Not only relatively simple perceptual phenomena seem to be subject to cultural relativity. Even well-known effects such as cognitive dissonance—the state of tension resulting from contradictory cognitions, as described by Leon Festinger (1957)—do not seem to be free from cultural influences (e.g., Aggarwal et al., 2013; Yakın et al., 2023). Even more convincing are the findings on the so-called *fundamental attribution error*. This is a judgment effect whereby people, when assessing the behavior of others, are more likely to attribute the behavior to the person, i.e., individual causes, rather than to the situation. Here, too, social psychology long assumed a universal phenomenon. By now, robust empirical evidence shows that the fundamental attribution error is more common in Western cultures than, for example, in Chinese, collectivist culture (e.g., Krull et al., 1999; Miyamoto & Kitayama, 2002). Even the tendency of individuals to prefer information that confirms their own attitudes—the so-called *confirmation bias*—is likely to be culture-dependent. At least, this is suggested by (non-representative) experimental comparative studies with U.S., German, and Japanese participants. People from all three cultures are not immune to preferring information that confirms their views. However, the confirmation bias is, unsurprisingly, particularly pronounced among Americans (Knobloch-Westerwick et al., 2019).

In the highly topical psychological research on dealing with climate change (see *Chap. 17*), the work also seems to come primarily from the "weird countries" (the Western, educated, industrialized, rich, and democratic societies). For example, Kim-Pong Tam, Angela Leung, and Susan Clayton analyzed 130 psychological studies on climate change and global warming in a systematic review. More than 92% of these studies were conducted in Western, democratic societies (Tam et al., 2021, p. 131). Only two studies involved participants from Africa, five from Latin America and the Caribbean, and eight studies included people from Asia.

What next? Methodological rigor, theoretical seriousness, and epistemological principles are important. Universally applicable rational principles for the "objective knowledge of reality" are not necessarily so. Certainly, and the experts know this as well, there are some ways to curb, for example, the publication crisis in psychology (and other disciplines): Replication studies should be encouraged, carefully justified, preferably conducted in cooperation with the research team of the study to be replicated, transparently documented, and published in an appropriate form (Ulrich et al., 2016, p. 171).

Moreover, it would be absurd and chauvinistic to claim that "Western" psychology has a monopoly on psychological knowledge. Because: "It is short-sighted to assume that one has 'solutions' for people whose lives one does not participate in and whose problems one does not know" (Feyerabend, 1980, p. 237). Psychologists should therefore advocate a pluralism of principles, theories, and methods in order to do justice to the possible diversity of realities and cultures. Cross-cultural studies and corresponding collaborations with colleagues from the Global South, for example, would be a step toward countering "Western" chauvinism.

14.4 Principles or Openness?—Reflections on the Methodological[5]

> "[…] there is no 'scientific method'; there is no single procedure, rule, there is no standard of excellence that applies to every research project and makes it 'scientific' and therefore trustworthy. Every project, every procedure, every theory must be measured on its own and by standards adapted to the relevant processes […] The idea of a universal and stable *method* and the corresponding idea of a universal and

[5] This section consists largely of self-citations (Frindte, 1998, pp. 263, Frindte & Geschke, 2019, pp. 77); freely following the motto: "[…] in the end, a good lecturer always notices when a text has been copied indiscriminately, and smells the forgery—whereas, I repeat, a well-chosen copy certainly deserves praise" (Eco, 2016, 48).

stable *rationality* are just as unrealistic as the idea of a measuring instrument that measures every quantity in all possible circumstances" (Feyerabend, 1980, p. 195; emphasis in original).

It would be absurd to rely solely on the rationality of psychological knowledge.

If everything that is said is said through a culturally shaped language, then the statements formulated within scientific communities about the psychological are also dependent on the characteristics of these language and interpretation communities.

Interpretive communities are social groupings of people who share the same or similar perspectives on social problems and processes, that is, largely inter-individually consistent social constructions. In the social science literature, one finds a number of terms intended to designate similar social realities as we attempt to do with the term *interpretive community*. Tilman Reitz (2014), following Ludwig Wittgenstein, speaks of "language communities," Zygmunt Bauman (1992) occasionally of "communities of meaning," and Ludwik Fleck of "thought collectives" (Fleck, 1993, original: 1935).

Every psychological statement is thus the result/construction of the rules of interpretation and communication as used by those scientists or scientific communities who have formulated these statements. Differences in psychological interpretations and communications therefore primarily express differences in the *social constructions* of the psychological. In this sense, every psychological statement is also a social artifact, a social work of art, a social construction, so I allow myself the following definition:

Psychological artifacts are social constructions about the psychological, fabricated by individual scientists and/or scientific communities, referring to those processes that can indeed be observed in individuals and, under certain circumstances, also self-reported by them, but are ultimately subject to the rules of interpretation and communication of the respective scientific community and must therefore be negotiated in public discourse with regard to their usefulness.

Psychological artifacts initially have one or more referents to which they relate, a *context of reference*. This is the domain of psychological phenomena that is to be explained and interpreted. The artifact research of the 1970s and 1980s illustrated in a differentiated way the problem of a psychological understanding according to which it would be possible to study the psychological processes of other people and obtain true statements about them; one would just have to find the right method. But such a method does not exist. Every researcher, if they wish to uncover the psychological regulatory mechanisms of their "subjects" (Ss), that is, other people—whether by experimental or non-experimental procedures—will be confronted with a serious control dilemma, which I would like to briefly outline as follows:

In almost any psychological investigation or research situation (in experiments as well as in narrative interviews, except perhaps for various non-reactive surveys), the researcher tries to confront the participants or Ss with certain action requirements (A) or "independent variables" (in the form of instructions, cover stories, interview introductions, etc.). If the researchers succeed in controlling that the participants or Ss base their individual reasoning for action solely on these requirements (A) (which is hardly to be expected, as artifact research has shown), the researcher is at least faced with the following question, which I call F1:

Are the actions (H) of the participants or Ss (e.g., behavior in the experiment, responses in the interview, etc.) actually expressions and indicators of their individual-specific organization of action, i.e., do the individual actions reveal something about the psychological particularities of the participants or Ss, or do the actions merely reflect the action requirements that are equally restrictive for all participants or Ss (deliberately set and controlled by the "experimenters")?

If, on the other hand, the researchers refrain from deliberately and systematically setting action requirements (A) and instead leave it to the participants and Ss to act as they wish (e.g., by using non-reactive procedures), and if the researcher succeeds in precisely controlling and recording the actions of the participants or Ss (which again is hardly conceivable), then the researchers are faced with another question, which I would like to call F2:

Which characteristics of the investigation situation (and other contextual conditions) did the participants or Ss consider in their individual reasoning for action as requirements to behave in this way and not otherwise; which action requirements, therefore, could I as a researcher take into account in the psychological interpretation of the participants' behavior?

Again, the researchers would face the problem of only being able to infer the characteristics of the psychological processes of their Ss vaguely. The control dilemma that arises when one tries to decide between the two questions (F1 and F2) is simply and plainly related to an uncertainty relation, which one always encounters when intending to study psychological processes. This uncertainty relation—by analogy with Heisenberg—could perhaps be simply symbolized as follows:

U (A) x U (H) > G (Psi),

where "U" stands for "uncertainty in the control of…", "A" refers to the action requirements or "independent variables" set by the "experimenters" (in the sense described above), "H" is meant to symbolize the actions produced by the participants or Ss and externally recorded, and "G (Psi)" denotes the "theoretically conceivable accuracy for capturing the individual psychological regulatory processes." According to this relation, reducing the uncertainty in capturing one variable (A or H) can only be achieved by increasing the uncertainty in capturing the other variable (H or A). Psychological processes are therefore hardly "objectively" accessible. This is simply and plainly due to the subjectivity of the psychological "object." Humans are self-aware, self-reflective, and capable of action, and, in cooperation with others, capable of self-emancipation.

This is a thesis of the philosophy of science, not one specific to psychology.

The psychology-specific aspect is somewhat more complicated. In (social) psychology, so-called dual-process models have been highly regarded since the 1980s; for example, the Elaboration Likelihood Model (ELM) by Petty and Cacioppo (1986), the Heuristic-Systematic Model (HSM) by Chaiken et al. (1989), or the MODE model by Fazio (1990); Kahneman's model of two cognitive processing systems (2012) can also be classified here. These models distinguish between two processes of information processing. In the ELM, they are called the peripheral and central routes; in the HSM, systematic processing is distinguished from heuristic processing; and in the MODE model, the distinction is between automatic or spontaneous versus planned or controlled processes; Kahneman refers to the two processes as System 1 and System 2. The central question these models seek to answer can be simply formulated as follows: When do we judge the world we are involved in more affectively, automatically, and according to internal schemas, and when do we undertake the effort to comprehend the world in a complex, conscious manner? John A. Bargh is not only credited with anchoring the unconscious in academic psychology (see also *Chap. 12*). He also attempted to newly explain the unconscious as an automatic process. An unconscious, automatic process can occur, for example, when a person is not aware of a stimulus (a stimulus from the environment is only perceived subliminally), when the person does not know how to interpret this stimulus or cannot do so, or when it is efficient (and must be done quickly) to react to such stimuli, or when the person is unable to stop their reactions to certain stimuli (Bargh, 2014).

An experiment may serve to illustrate (subliminal perception). Although it is not that old, it has already become a classic when it comes to the automatic processing of stimuli, and it originates from Bargh et al. (1996). It is a priming experiment. Thirty students participated in the experiment, which was presented as a language test. They were given a word list containing 30 groups of words, each with five word variations. From these, participants were to form a grammatically correct sentence with four words as quickly as possible. The word group lists contained either neutral words or words that could be associated with the stereotype *elderly people*, e.g., old, lonely, gray, forgetful, dependent, helpless, etc. Some participants worked on the neutral list, while the others worked on the list with age-stereotypical words. After the experiment, the participants left the experimental room via the hallway toward the elevator. In the hallway sat an assistant of the experimenter, who was not recognizable as such to the participants. The assistant recorded how long it took each participant to walk from the experimental room to the elevator. The results: The participants who had previously worked on the list with age-stereotypical words took more time to reach the elevator than those who had been given the neutral word list.

The authors interpret the results as follows: For those participants who worked on a list with age-stereotypical words, an age stereotype was unconsciously and automatically activated, leading to motor behavior compatible with the activated stereotype—walking slowly and leisurely. After the experiment, the participants stated that they had not noticed that the words used in the experiment had a common theme.

Three conclusions may suffice: First, dual-process models of information processing suggest that there are automatic or unconscious processes, indicating that people do not always and everywhere appear to be self-reflective. Second, methods of introspection are hardly suitable for elucidating unconscious processes in dealing with the world. Third, and this seems to me the most important: Priming experiments or similar experimental designs do not circumvent the aforementioned control dilemma. Was it really the activation of age stereotypes that caused the primed participants to walk more slowly? Or did the participants walk more slowly because the age-stereotypical words prompted them to think about their recently deceased or ill grandparents? It is well known that some people tend to walk more slowly when they are consciously searching for the solution to a problem. Perhaps the participants did in fact recognize the actual aim of the experiment and, as "good subjects," wanted to do the experimenters a favor? Or are other interpretations conceivable? "No thought is so old or absurd that it could not improve our knowledge" (Feyerabend, 1986, p. 55).

Psychological processes are explained by psychological artifacts, that is, by socially fabricated constructs. The social domain in which this fabrication takes place could also be called the *Context of Manufacture*. These are the *interaction spaces* in which psychologists act, experiment, discuss, and reflect on their psychological activities. In this social domain, *psychological theories* are invented and discovered, *psychological data* are observed, and *social constructions about the psychological* are fabricated. Let us first ask: How does one find a theory?

The social psychologists Elliot Aronson and Merrill Carlsmith wrote in 1968: "[…] where the ideas come from is not terribly important […] the important and difficult feat involves translating a conceptual notion into a tight, workable, credible, meaningful set of experimental operations" (Aronson & Carlsmith, 1968, p. 37; quoted in Kruglanski, 2001, p. 871). The authors address a distinction that has been known and named in the philosophy of science at least since Hans Reichenbach: the distinction between *Context of Discovery* and *Context of Justification* (Reichenbach, 1938). In contrast to Reichenbach, Paul Feyerabend famously argued that both contexts are permeated with subjective elements (see, e.g., Feyerabend, 1992, pp. 157).

Chance, inspiration, intuition, luck, or speculative imagination seem to influence both the development of theories and the appropriate choice of methods to test them. Or, as Kruglanski puts it, "inspiration, intuition, and imagination" (Kruglanski, 2001, p. 874).

A few years earlier, Gerd Gigerenzer answered the question of how one finds a theory as follows: "In the English-speaking world, it is said that scientific theories are born in the context of the three 'B's': 'bed', 'bathroom', and 'bicycle' […] Textbooks and curricula teach which methods can be used to find out whether theories are right or wrong and under what conditions; but little about where these theories come from. […] the focus is mainly on the 'context of justification', i.e., the context in which existing theories are tested; the 'context of discovery', i.e., the context in which theories arise, remains in the dark" (Gigerenzer, 1988, p. 91). Gigerenzer primarily blames Karl R. Popper for this obscurity in the *context of discovery*.

I would add: Paul Feyerabend seems to see it similarly. Philosophers like Popper have fostered this ignorance by portraying the emergence of theories as ultimately inexplicable, mystical events. Perhaps that is the case. The *Critical Rationalism* of Popperian origin still seems to be the implicit, subjective philosophy of science for many social scientists. At least, such a suspicion is not lessened by the explicit identification with critical rationalist views by German social psychologists (see, e.g., Frey & Bierhoff, 2011, pp. 285).

So once again: Where do (social-)psychological theories come from? Gerd Gigerenzer's answers to this question were already published in 1987 (Gigerenzer & Murray, 1987), later expanded and slightly modified by him (1994), and have since been frequently cited.

"To the question of the origin of theories, one can [...] at least receive three classes of answers: 1. Stories, such as that G.T. Fechner, lying in bed on October 22, 1850, suddenly envisioned the concept of the psychophysical function; 2. the inductive argument that new theories arise from new data; and 3. the argument that new theories are motivated by new metaphors" (Gigerenzer, 1994, p. 111).

Gigerenzer's attention is directed to the third argument.

"Since the so-called cognitive turn around 1960 (when cognitive psychology became the central psychological paradigm; WF), many theories have been based on metaphors drawn from the 'toolbox' of the social scientist. The two most important tools that became central to theories about cognitive processes are *statistics* and the *computer*. Originally, both were merely means for analyzing data, but by now they have shaped the vocabulary of theories since the cognitive turn [...] This suggests taking the researcher's activity, which is most familiar to us, as a metaphor for that unknown which takes place in the minds of the subjects under study" (Gigerenzer, ibid., p. 113).

Put pointedly, one could claim: Psychologists in general, and social psychologists in particular, construct theories in the image they have of their own tools and of themselves as tool-making and tool-using subjects. The engineer who has become a psychologist regards his psychological subject matter as a well-structured machine. The attribution researcher familiar with analysis of variance, as for example Harold Kelley (1967) did, designs a theory of causal attributions in analogy to precisely this analysis of variance. And so on and so forth. The process of theory formation in the research domain could proceed in this or a similar way.

But since this context of knowledge fabrication is always also a social one, in which psychologists usually work in groups of scientists and are in social exchange with other groups and communities, we must push the above point even further: Psychologists fabricate knowledge against the background of the interaction and communication patterns with which they are accustomed to constructing their own realities. In other words: The interpretations of the world that are shared and passed on interindividually within the respective groups, including the image of their own group, function as important metaphors for their own theory formation. The way in which, for example, group decisions are made in such groups, how groupthink effects develop, how minorities are included or not included in the joint research process, how one can count on state funding, how myths about scientific expertise are cultivated and spread—all this and more belongs to the bubbling source from which psychologists draw their theories about non-psychologists.

But even more happens in the research context: Against the background of more or less elaborate theories, psychologists fabricate their knowledge by observing, experimentally influencing, or questioning reference persons, subjects, or experimental groups. That is: The research context is not only the social space in which psychologists (and other scientists) attempt to discover their theories about psychologically relevant phenomena. It is also the space in which the empirical data are fabricated, with which the theories are justified and which itself is to be explained by the theories. In this context, the carousel of theory and empiricism ultimately turns. And this carousel turns as a result of an inner dynamic: To observe means to make a distinction and at the same time to designate one side of the distinction, to name it, to link it with a concept, in order to better grasp it in the future. The side of the distinction that is excluded from observation virtually disappears from view, does not need to be explained, and is not used to justify explanations.

Scientific observers see only what they consider worthy and capable of explanation, and they explain, with the help of their theories, only what they believe they see and observe through the lens of their theory.

In the language of philosophers of science, we are dealing here *on the one hand* with the well-known problem of the theory-ladenness of observation, which Paul Feyerabend has explained at length. *On the other hand*—and this is likely decisive—the carousel of theory and empiricism generated by actors in the research context is a self-referential, that is, self-related, cycle.

Is it not the case that scientists, although they know and master the rules of good scientific practice (e.g., formulating research questions and justifying hypotheses, selecting appropriate research methods, quantitative and qualitative data analysis and interpretation), in the actual research process and in the interest of knowledge, often violate and must violate these rules? I do not mean the already mentioned and also present in (social) psychology conscious deception and lying by some "black sheep." No, I mean the way in which theory and empiricism are related to each other. Theory and empiricism are often varied until they fit together and allow for a relatively stable interpretation of the findings. This approach is not only common; it is also sensible. For the usefulness of the results does not depend on rigidly prescribed rules of how research should be conducted, but must ultimately prove itself in 'real life.'

For the fact that researchers fabricate their knowledge in such a recursive research process, they must pay a high price; the price of being struck by blindness in two ways: On the one hand, by deciding to observe and investigate something, researchers exclude those other possibilities that could also potentially be observed. On the other hand, the distinction that researchers have made in order to observe something is always concealed. The decision once made to observe something is no longer present in the process of observation and investigation, for no observer can observe his or her own observation while observing (cf. Luhmann, 1991, p. 63). However: Another observer can observe which distinctions the observed observers (i.e., our researchers) use, and thus see what they cannot see. With such observation of the *second* order, it becomes possible to make present again those sides that first-order observers had previously excluded from their observation. That is, second-order observation could help to uncover the blind spots of first-order observers. This is precisely what happens in a context I like to call the *Context of Validation* (validation space). These are the social domains in which scientific knowledge attains its actual scientific and practical validity, by being legitimized by the *scientific community* and accepted as genuinely relevant by other social groups. Those who have fabricated their knowledge about the psychological in the research context must—if they want their fabrications to be recognized as valid—demonstrate that the knowledge fits into the scientifically recognized spaces of meaning. In other words: The actors must demonstrate that they interpret psychological realities in a way that corresponds to the interindividually consistent conceptions of the psychological in scientifically established *interpretive communities.*

Years ago, Jürgen Klüver (1988) called the attempts by which scientists try to explicate and validate their scientific fabrications *literary* and *collective validation.* The aim is to integrate the "true" scientific knowledge fabricated in the "ivory tower" or in the research group into the space of meaning of one's own scientific community, that is, to practice follow-up action by aligning one's own scientific products with the community's body of knowledge until they fit. The scientific community itself thereby assumes the function of the second-order observer. And these second-order observers generally demand that those who wish to have their "fabrications" validated observe the conventionalized and traditional rules of scientific interpretation and communication.

But, as Paul Feyerabend notes: "*You simply cannot rely on scientists.* They have their own interests, which color their interpretation of the evidence and the cogency of that evidence, they *know* very little, but pretend to know much more, they use rumors as if they were well-established facts, pious wishes as if they were fundamental 'principles' of scientific thinking, and even the very detailed research results are based on assumptions that the scientists themselves often do not know and whose content and scope they do not understand..." (1980, pp. 188; emphasis in the original).

And so it is not surprising when scientists seeking validation, for whom their own publicity is more important than the seriousness of their scientific fabrications, attempt to conduct scientific discourse according to rules of communication that run counter to (supposedly) scientific rationality. *Publish or perish.*

References

Aggarwal, P., Soo Kim, C., & Cha, T. (2013). Preference-inconsistent information and cognitive discomfort: A cross-cultural investigation. *Journal of Consumer Marketing, 30*(5), 392–399.

Arnett, J. (2008). The neglected 95%: Why American psychology needs to become less American. *American Psychologist, 63*(7), 602–614.

Aronson, E., & Carlsmith, J. M. (1968). Experimentation in social psychology. In G. Lindzey & E. Aronson (Hrsg.), *Handbook of Socialpsychology* (Vol. 2, pp. 1–79). Addison Wesley.

Ash, M. G. (1985). Die experimentelle Psychologie an den deutschsprachigen Universitäten von der Wilhelminischen Zeit bis zum Nationalsozialismus. In M. G. Ash & U. Geuter (eds.), *Geschichte der deutschen Psychologie im 20. Jahrhundert* (pp. 45–82). Westdeutscher Verlag.

Ash, M. G. (2023). Psychology and the Fall of Communism: The special case of (East) Germany. *Journal of the History of the Behavioral Sciences, 59*(1), 8–19.

Bak, P. M. Zum Verhältnis von Neurowissenschaft und Psychologie. *Journal of Business and Media Psychology, 11*, 1, pp. 1–11.

Bargh, J. (2017). *Before you know it. The unconscious reasons we do what we do*. Simon & Schuster.

Bargh, J. A. (2014, second Edition). The four horsemen of automaticity: Awareness, intention, efficiency, and control in social cognition. In R. S. Wyer & T. K. Srull (eds.), *Handbook of Social Cognition, Vol. 1.* (pp. 1–40). Psychology Press.

Bargh, J. A., Chen, M., & Burrows, L. (1996). Automaticity of social behavior: Direct effects of trait construct and stereotype activation on action. *Journal of personality and social psychology, 71*(2), 230.

Bauman, Z. (1992). *Moderne und Ambivalenz*. Junius.

Bohner, G. (2016). Experimentalpsychologische Genderforschung in Bielefeld. *Onlinezeitschrift des Interdisziplinären Zentrums für Geschlechterforschung (IZG)*, pp. 58–62.

Brachem, J., Frank, M., Kvetnaya, T., Schramm, L. F., & Volz, L. (2022). Replikationskrise, p-hacking und Open Science. *Psychologische Rundschau, 73*(1), 1–17.

Brand, H.W. (1980). “Soziale Wahrnehmung”—oder Wahrnehmung in sozialen Situationen. In W. Bungard (ed.), *Die “gute” Versuchsperson denkt nicht*. Urban & Schwarzenberg.

Bredenkamp, J. (1993). Zur Lage der Psychologie in den neuen Bundesländern. *Psychologische Rundschau, 44*, 1–10.

Breyer, T. (2017). Soziale Wahrnehmung zwischen Erkenntnistheorie und Anthropologie. *Interdisziplinäre Anthropologie: Jahrbuch, 4*(2016), 141–161.

Bühler, K. (1908). Antwort auf die von Wundt erhobenen Einwände gegen die Methode der Selbstbeobachtung an experimentell erzeugten Erlebnissen. *Archiv für die gesamte Psychologie, 12*, 93–122.

Bühler, K. (1927). *Die Krise der Psychologie*. Barth.

Bungard, W. (1984). *Sozialpsychologie im Labor*. Göttingen: Verlag für Psychologie C.J. Hogrefe.

Busse, S. (1993). Gab es eine DDR-Psychologie?. *Psychologie und Geschichte*, *5*(1/2).
Chaiken, S., Liberman, A., & Eagly, A. H. (1989). Heuristic and systematic information processing within and beyond the persuasion context. In J. S. Uleman & J. A. Bargh (eds.), *Unintended thought* (pp. 212–252). Guilford.
Diekmann, A. (2021). Unehrliche Ehrlichkeitsforschung. *Frankfurter Allgemeine Zeitung* vom 8. September 2021.
Dilthey, W. (1894/1990). Ideen über eine beschreibende und zergliedernde Psychologie. In G. Misch (ed.), *Gesammelte Schriften* (vol. 5, pp. 139–240). Teubner Verlagsgesellschaft.
Ebbinghaus, H. (1896). Über erklärende und beschreibende Psychologie. *Zeitschrift für Psychologie und Physiologie der Sinnesorgane, 9*, 161–205.
Eco, U. (2012). *Im Krebsgang voran*. Deutscher Taschenbuch Verlag.
Eco, Umberto (2016). *Pape Satàn*. Hanser.
Eyssel, J., Geschke, D., & Frindte, W. (2015). Is seeing believing? The relationship between TV consumption and Islamophobia in German majority society. *Journal of Media Psychology: Theories, Methods, and Applications, 27*(4), 190.
Fanelli, D. (2009). How many scientists fabricate and falsify research? A systematic review and meta-analysis of survey data. *PLoS ONE, 4*(5), e5738.
Fazio, R. H. J. (1990). Multiple processes by which attitudes guide behaviour: The MODE model as an integrative framework. In M. P. Zanna (ed.), *Advances in experimental social psychology* (vol. 23, pp. 75–109). Academic.
Festinger, L. (1957). *A theory of cognitive dissonance*. Stanford University Press.
Feyerabend, P. K., & Thomas, C. (Hrsg.). (1986). *Nutzniesser und Betroffene von Wissenschaften*. Verlag der Fachvereine.
Feyerabend, P. K. (1980). *Erkenntnis für freie Menschen*. Suhrkamp.
Feyerabend, P. K. (1986; Original: 1975). *Wider den Methodenzwang*. Suhrkamp.
Feyerabend, P. K. (1992; Original: 1989). *Über Erkenntnis. Zwei Dialoge*. Campus.
Feyerabend, P. K. (2005). *Die Vernichtung der Vielfalt. Ein Bericht*, herausgegeben von Peter Engelmann. Passagen Verlag.
Fleck, L. (1993, Original: 1935). *Entstehung und Entwicklung einer wissenschaftlichen Tatsache*. Suhrkamp.
Frey, D. & Bierhoff, H.-W. (2011). *Sozialpsychologie—Interaktion und Gruppe*. Göttingen u.a.: Hogrefe.
Frindte, W. & Geschke, D. (2019). *Lehrbuch Kommunikationspsychologie*. Beltz/Juventa.
Frindte, W. & Haußecker, N. (2010). *Inszenierter Terrorismus*. Springer VS.
Frindte, W. (1998). *Soziale Konstruktionen*. Westdeutscher.
Frindte, W. (2022). *Quo Vadis, Humanismus?* Springer.
Galliker, M. (2016). *Ist die Psychologie eine Wissenschaft?* Springer VS.
Gergen, K. J. (1989). Social psychology and the wrong revolution. *European Journal of Social Psychology, 19*(5), 463–484.
Gigerenzer, G., & Murray, D. J. (1987). *Cognition as intuitive statistics*. Erlbaum.
Gigerenzer, G. (1988). Woher kommen Theorien über kognitive Prozesse? *Psychologische Rundschau, 39*, 91–100.
Gigerenzer, G. (1994). Woher kommen die Theorien über kognitive Prozesse? In A. Schorr (ed.). *Die Psychologie und die Methodenfrage*. Hogrefe.

Goethe, J. W. (1972). Maximen und Reflexionen. *Berliner Ausgabe, Band 18*. Aufbau Verlag.

Greenwood, J. D. (1983). Role-playing as an experimental strategy in social psychology. *European Journal of Social Psychology, 13*(3), 235-254.

Groeben, N. & Scheele, B. (1977). *Argumente für eine Psychologie des reflexiven Subjekts*. Steinkopff.

Gummersbach, W. (1985). Krise der Psychologie. Zur Aktualität eines traditionellen Themas. In M. G. Ash & U. Geuter (eds.), *Geschichte der deutschen Psychologie im 20. Jahrhundert* (pp. 314–339). Westdeutscher.

Hacker, W. (1973). *Allgemeine Arbeitspsychologie*. Psychische Struktur und Regulation von Arbeitstätigkeiten: Deutscher Verlag der Wissenschaften.

Hacker, W. (2021). *Psychische Regulation von Arbeitstätigkeiten 4.0*. Zürich: Hochschulverlag AG.

Henrich, J., Heine, S. J., & Norenzayan, A. (2010). The weirdest people in the world? *Behavioral and brain sciences, 33*(2–3), 61–83.

Herrmann, T. (1979). *Psychologie als Problem*. Klett-Cotta.

Th., & Herrmann. (1991). Diesmal diskursiv—schon wieder eine Erneuerung der Psychologie. *Report Psychologie, 2*, 21–27.

Herzog, W. (2012). *Wissenschaftstheoretische Grundlagen der Psychologie*. Springer VS.

Hiebsch, H. & Vorwerg, M. (eds.) (1979). *Sozialpsychologie*. Deutscher Verlag der Wissenschaften.

Hiebsch, H. (1977). *Wissenschaftspsychologie*. Deutscher Verlag der Wissenschaften.

Hiebsch, H. (1986). *Interpersonelle Wahrnehmung und Urteilsbildung*. Deutscher Verlag der Wissenschaften.

Holzkamp, K. (1973). *Sinnliche Erkenntnis*. Fischer Athenäum Taschenbuch.

Holzkamp, K. (1983). *Grundlegung der Psychologie*. Campus.

Jaeger, S. (1993). Zur Widerständigkeit der Hochschullehrer zu Beginn der nationalsozialistischen Herrschaft. *Psychologie und Geschichte, 4*, 219–228.

Kahneman, D. (2011). *Thinking, Fast and Slow*. Farrar, Straus and Giroux.

Kamin, L. J. (1979; Original: 1974). *Der Intelligenz-Quotient in Wissenschaft und Politik*. Steinkopf.

Kelley, H. H. (1967). Attribution theory in social psychology. In D. Levine (Ed.), *Nebraska Symposium on Motivation* (Vol. 15, pp. 129–238). University of Nebraska Press.

Kelman, H. C. (1967). Human use of human subjects: The problem of deception in social psychological experiments. *Psychological Bulletin, 67*(1), 1–11.

King, E. B., Avery, D. R., Hebl, M. R., & Cortina, J. M. (2018). Systematic subjectivity: How subtle biases infect the scholarship review process. *Journal of Management, 44*(3), 843–853.

Klix, F. (1971). *Information und Verhalten*. Kybernetische Aspekte der organismischen Informationsverarbeitung: Deutscher Verlag der Wissenschaften.

Klix, F. (Hrsg.). (1984). *Gedächtnis, Wissen*. Wissensnutzung: Deutscher Verlag der Wissenschaften.

Klüver, J. (1988). *Die Konstruktion der sozialen Realität Wissenschaft: Alltag und System*. Vieweg.

Knobloch-Westerwick, S., Liu, L., Hino, A., Westerwick, A., & Johnson, B. K. (2019). Context impacts on confirmation bias: Evidence from the 2017 Japanese snap election compared with American and German findings. *Human Communication Research, 45*(4), 427–449.

Konitzer, F. (2013). Einmal ist keinmal. *Bild der Wissenschaft, 8*, 66–69.

Kriz, J. (1981). *Methodenkritik empirischer Sozialforschung*. B.G: Teubner.

Kruglanski, A. W. (2001). That "Vision Thing". The State of Theory in Social and Personality Psychology at the Edge of the New Millennium. *Journal of Personality and Social Psychology, 80*(6), pp. 871–875.

Krull, D. S., Loy, M. H. M., Lin, J., Wang, C. F., Chen, S., & Zhao, X. (1999). The fundamental fundamental attribution error: Correspondence bias in individualist and collectivist cultures. *Personality and Social Psychology Bulletin, 25*(10), 1208–1219.

Legewie, H. (1991). Argumente für eine Erneuerung der Psychologie. *Report Psychologie, 2*, 11–20.

Lenski, K. (2021). Die Sozialpsychologie der DDR und die Staatssicherheit. Örtliche und überregionale Verflechtungen am Beispiel der FSU Jena. In A. Maercker & J. Giesecke (Hrsg.), *Psychologie als Instrument der SED-Diktatur*. Hogrefe.

Luhmann, N. (1991). Wie lassen sich latente Strukturen beobachten? In P. Watzlawick & P. Krieg (eds.), *Das Auge des Betrachters*. Piper.

Maercker, A. & Gieseke, J. (Hrsg.) (2021). *Psychologie als Instrument der SED-Diktatur*. Hogrefe.

Maiers, W. (2021). Psychologie in der Replikationskrise—eine Replikation ihrer Krisen? In M. Dietrich, I. Leser, K. Mruck, P. S. Ruppel, A. Schwentesius, & R. Vock (eds.), *Begegnen, Bewegen und Synergien stiften: Transdisziplinäre Beiträge zu Kulturen, Performanzen und Methoden* (pp. 441–459). Springer VS.

Mann, Th. (1954). *Doktor Faustus*. Aufbau Verlag.

Markard, M. (1985). Konzepte der methodischen Entwicklung des Projekts Subjektentwicklung in der frühen Kindheit. *Forum Kritische Psychologie, 17*, 101–125.

Maschewsky, W. (1977). *Das Experiment in der Psychologie*. Campus.

Massen, C., & Bredenkamp, J. (2005). Die Wundt-Bühler-Kontroverse aus der Sicht der heutigen kognitiven Psychologie. *Zeitschrift für Psychologie, 213*(2), 109–114.

Mayring, P. (1990). *Einführung in die qualitative Sozialforschung*. Psychologie Verlagsunion.

Mertens, W. & Fuchs, G. (1978). *Krise der Sozialpsychologie?* Ehrenwirth.

Mertens, W. (1980). Artefakte in der Aggressionsforschung. In W. Bungard (Hrsg.), *Die „gute" Versuchsperson denkt nicht. München.* Urban & Schwarzenberg.

Merton, R. K. (1957). *Social theory and social structure*. Free Press.

Métraux, A. (1985). Der Methodenstreit und die Amerikanisierung der Psychologie in der Bundesrepublik 1950–1970. In G. M. Ash & U. Geuter (eds.), *Geschichte der deutschen Psychologie im 20. Jahrhundert*. (pp. 225–251). Westdeutscher Verlag.

Miyamoto, Y., & Kitayama, S. (2002). Cultural variation in correspondence bias: The critical role of attitude diagnosticity of socially constrained behavior. *Journal of Personality and Social Psychology, 83*(5), 1239–1248.

Open Science Collaboration. (2015). Estimating the reproducibility of psychological science. *Science, 349*(6251), aac4716.

Orne, M. T. (1962). On the social psychology of the psychological experiment: With particular reference to demand characteristics and their implications. *American Psychologist, 17*(11), 776–783.

Peter, C., & Brosius, H.-B. (2013). Wahrnehmungsphänomene. In W. Schweiger & A. Fahr (Hrsg.), *Handbuch Medienwirkungsforschung* (pp. 463–480). Springer VS.

Petty, R. E. & Cacioppo, J. T. (1986). The Elaboration Likelihood Model of persuasion. In L. Berkowitz (ed.). *Advances in Experimental Social Psychology* (vol. 19, pp. 123–205). Academic Press.

Pforr, K., & Schröder, J. (2015). Warum Panelstudien? (Version 1.1). (GESIS Survey Guidelines). Mannheim: GESIS -Leibniz-Institut für Sozialwissenschaften. https://doi.org/10.15465/gesis-sg_00.

Reichenbach, H. (1938). *Experience and Prediction.* An Analysis of the Foundations and the Structure of Knowledge: The University of Chicago Press.

Reitz, T. (2014). *Sprachgemeinschaft im Streit. Philosophische Analysen zum politischen Sprachgebrauch.* transcript.

Rost, D. H., & Bienefeld, M. (2019). Nicht replizieren: Publizieren!? *Zeitschrift für Pädagogische Psychologie, 33*(4–3), 163–176.

Sabel, B. A., Knaack, E., Gigerenzer, G., & Bilc, M. (2023). Fake Publications in Biomedical Science: Red-flagging Method Indicates Mass Production. *medRxiv*, 2023–05; https://doi.org/10.1101/2023.05.06.23289563. Accessed: June 22, 2023.

Schreier, M., Odağ, Ö. (2020). Mixed Methods. In G. Mey & K. Mruck (eds.),) *Handbuch Qualitative Forschung in der Psychologie.* (pp. 159–184). Springer.

Schuler, H. (1980). *Ethische Probleme psychologischer Forschung*. Hogrefe.

Segall, M. H., Campbell, D. T. & Herskovits, M. J. (1966) *The Influence of Culture on Visual Perception.* Bobbs-Merrill.

Semin, G. R., & Fiedler, K. (1988). The cognitive functions of linguistic categories in describing persons: Social cognition and language. *Journal of Personality and Social Psychology, 54*, 558–568.

Shu, L. L., Mazar, N., Gino, F., Ariely, D., & Bazerman, M. H. (2012). Signing at the beginning makes ethics salient and decreases dishonest self-reports in comparison to signing at the end. *Proceedings of the National Academy of Sciences, 109*(38), 15197–15200.

Stern, W. (1913). *Die Anwendung der Psychoanalyse auf Kindheit und Jugend.* Ein Protest: Verlag Johann Ambrosius Barth.

Strack, F. (1988). Social Cognition: Sozialpsychologie innerhalb des Paradigmas der Informationsverarbeitung. *Psychologische Rundschau, 39*(2), 72–82.

Tam, K. P., Leung, A. K. Y., & Clayton, S. (2021). Research on climate change in social psychology publications: A systematic review. *Asian Journal of Social Psychology, 24*(2), 117–143.

Timaeus, E. (1974). *Experiment und Psychologie*. Zur Sozialpsychologie psychologischen Experimentierens: Hogrefe.

Ulrich, R., Erdfelder, E., Deutsch, R., Strauß, B., Brüggemann, A., Hannover, B., & Rief, W. (2016). Inflation von falsch-positiven Befunden in der psychologischen Forschung. *Psychologische Rundschau, 67*(3), 163–174.

Volmerg, B. (1992). Zur Gründung der Neuen Gesellschaft für Psychologie (NGfP). *Journal für Psychologie, 1*(1), 36–42.

Walach, H. (2020). *Psychologie: Wissenschaftstheorie, philosophische Grundlagen und Geschichte*. Ein Lehrbuch: Kohlhammer.

Webb, E.J. (1975). *Nichtreaktive Messverfahren*. Beltz.

White, D. M. (1950). The "Gatekeeper". A Case Study in the Selection of News. *Journalism Quaterly, 27*, 383–390.

Wolfradt, U., Billmann-Mahecha, E., & Stock, A. (2017). *Deutschsprachige Psychologinnen und Psychologen 1933–1945*. Springer.

Woodward, R. W., & Clark, S. C. (1996). The Reflection of Soviet Psychology in East German Psychological Practice. In V. A. Koltsova, Y. N. Oleinik, A. R. Gilgen, & C. K Gilgen (eds.), *Post-Soviet Perspectives on Russian Psychology* (pp. 236–250). Greenwood Press.

Wundt, W. (1907). Über Ausfrageexperimente und über die Methoden zur Psychologie des Denkens. *Psychologische Studien, 3*, 301–306.

Yakın, V., Güven, H., David, S., Güven, E., Bărbuţă-Mişu, N., Güven, E. T. A., & Virlanuta, F. O. (2023). The Effect of Cognitive Dissonance Theory and Brand Loyalty on Consumer Complaint Behaviors: A Cross-Cultural Study. *Sustainability*, *15*(6), pp. 4718 ff. https://doi.org/10.3390/su15064718.

"Language that Writes Poetry and Thinks for You"—White Men and "Cancel Culture"

15

"If the new *language* spreads, it affects the articulation of the life of the soul and thus of perception, then soon the human being finds himself in a new environment, he perceives new objects, he lives in a new world" (Feyerabend, 2009, p. 169; emphasis in the original).

15.1 Language Games

"[…] Language that composes poetry and thinks for you," says a verse by Friedrich Schiller. Victor Klemperer, the Romance philologist who authored the "LTI-Notizbuch eines Philologen," picks up Schiller's line and writes: "But language not only composes poetry and thinks for me, it also guides my feelings, it steers my entire spiritual being, the more naturally, the more unconsciously I surrender myself to it. And what if the cultivated language has been formed from poisonous elements or has been made the carrier of toxins? Words can be like tiny doses of arsenic: they are swallowed unnoticed, they seem to have no effect, and after some time the toxic effect is still there" (Klemperer, 1970, p. 24; original: 1947).

It is about the influence of language on our thinking and about our thinking, which seeks language. These are questions that Paul Feyerabend also grappled with throughout his life (e.g., with the relationship between observational and everyday languages or with the relations between language and theory, Feyerabend, 1958, 1970, 1987).

W. Frindte, *Against Narrow-Mindedness and Chauvinism—Journeying Through Absurd Times with Paul K. Feyerabend*,
https://doi.org/10.1007/978-3-658-50113-6_15

> "The successful researcher is an educated person, he knows many tricks, ideas, ways of speaking, he knows details of the history of his field as well as cosmological abstractions, rumors as well as facts, he can combine fragments from very different standpoints and quickly move from one framework to a completely different and incommensurable one. He is not bound to any particular language, he sometimes speaks the language of facts, then again the language of fairy tales and mixes them in very unexpected ways. And note that such procedures appear both in the 'context of discovery' and in the 'context of justification,' for the scientific *investigation* of ideas is just as complex as their *invention*" (Feyerabend, 1981, pp. 55; emphasis in the original).

With the words that have been found or invented, what is seen can be named. What is seen, which has been created or constructed, receives a name. With the words it becomes real, because now what has an effect can also be recognized and named. It enters the world and can be distinguished from other things in this world. With the words, what has been created, constructed, seen, and named begins to unfold its effect in the world. So far, so general.

It becomes concrete when it comes to the political aspect of language, its use, and the criticism of seemingly traditional designations, words, phrases, formulations—in other words, *language games.*

It is well known that social conceptions or constructions of reality in social communities (societies, organizations, groups, etc.) are interpretations of the world—including the world of one's own community—that are shared (conventionalized) and passed on (traditionalized) by the respective members. When they are expressed linguistically, externalized, this usually happens in various language games and according to the rules of these language games. To recall the Wittgenstein program: With the word "language game," Wittgenstein wants to emphasize "that speaking a language is part of an activity, or of a form of life" (Wittgenstein, 1984, Philosophical Investigations, Part I, § 23). The word "apple" can, within certain forms of thought, language, and life, hardly—as Wittgenstein describes (Wittgenstein, 1988, Remarks on the Philosophy of Psychology, note 489)—be replaced by the word "bank." Such a substitution violates the rules, conventions, and traditions within certain social communities. That is, the contents of social constructions, what they ultimately refer to, these contents apparently resist arbitrary changes to a certain extent. And the members of social communities often do so as well. Some are reluctant to give up "Gypsy schnitzel," still want to go to the "Moor's pharmacy" they trust, play "Indians," experience Shakespeare's "Othello," the "Moor of Venice," in the theater with good actors, continue to live on Nettelbeckufer, think Patti Smith's "Rock N Roll Nigger" is great, want to be able to call Immanuel Kant a great Enlightenment thinker, and not use gender asterisks or internal capitals, and so on and so forth.

15.2 On Recalcitrant Language Traditions

Let us look at some examples in order to subsequently review scientific findings that speak against recalcitrant language traditions. For one must not overlook, "that human beings can not only use languages, but also invent them," writes Feyerabend in another, but also fitting, context (Feyerabend, 1981, p. 90; original: 1962).

I will begin with the seemingly more problematic issue, because it is currently stirring up so much dust in political debates: gendering, that is, gender-fair speaking and writing. Politicians, such as the Minister President of Bavaria, AfD-affiliated citizens, journalists, but also scientists, speak and write of "gender ideology" and "gender madness." Some reject "Newspeak," "supervised speech" (Gauck, 2019) or cite understandable reasons to end gendering (Payr, 2021). The Ministry of Education in Saxony-Anhalt banned gendering with special characters in schools in mid-2023, i.e., asterisks, internal capitals, underscores, etc. (MDR, 2023). Others refer to the *Council for German Orthography,* which in March 2021 "did not recommend the inclusion of asterisk ('gender star'), underscore ('gender gap'), colon or other abbreviated forms to denote multi-gendered terms within words in the official rules of German orthography at this time" (Council for German Orthography, 2021). Still others point to the numerous perception studies showing that the generic masculine (the person or occupational designation in grammatically masculine form) in its application (e.g., doctor or psychologist) predominantly triggers male images and schemas (e.g., Nübling, 2018). Empirical studies can also refute concerns that gender-fair formulations (e.g., doctor and female doctor, psychologist and female psychologist) would indirectly reinforce gender discrimination (Cirksena & Leiner, 2022). There are, however, indications that gender-fairly formulated texts, in which, for example, the gender asterisk or internal capital is used, could reduce reading speed (Pöschko & Prieler, 2018).

Research that can clearly demonstrate whether and to what extent the various strategies of gender-inclusive language and writing (e.g., the gender asterisk or the capital "I" within words) can promote cognitive differentiation of gender diversity is relatively rare. And yet, there are some empirically convincing findings suggesting that the gender asterisk and the underscore are more likely than the capital "I" or dual forms (e.g., "female and male colleagues") to foster more inclusive mental representations (e.g., Bröder et al. 2022; Löhr, 2021). [1] However, it would be absurd to believe that the gender asterisk could, in itself, eliminate discrimination based on gender.

The same probably applies to other terms and their renaming. Take the term "Gypsy." This designation has appeared since the 15th century in various European countries in different spellings, for example as "Cygan" in Polish, "Cyganki" in Russian, "Tsiganes" in French, "Tziganu" in Romanian, "Zingaro" or "Zingano" in Italian (Engbring-Romang, 2014; Schwicker, 1883; Wippermann, 1998). The origin of the term seems to be disputed. What is certain is that it is a negatively connoted exonym, which in German as "Zigeuner" probably first appeared in the late 14th and early 15th centuries.

Under National Socialism, the Sinti and Roma—as they call themselves—were listed in identification papers and in concentration camp records as "Z." Such camps existed, to "combat the Gypsy problem," in several regions of Germany,[2] for example in Berlin Marzahn, Düsseldorf, Essen, Frankfurt am Main, Fulda, Hanover, Kiel, Cologne, or Magdeburg. In fact, these were precursors to extermination (Milton, 1995). The extermination began on Himmler's orders in 1941. Sinti and Roma were sent to the extermination camps Chelmno or Auschwitz-Birkenau. In Auschwitz-Birkenau, over 23,000 Sinti and Roma were interned in the so-called "Gypsy camp" between 1943 and 1944; more than 19,000 died or were murdered. In Croatia, between 10,000 and 40,000 Sinti and Roma were killed; in Romania, around 25,000 by 1942. It is estimated that

---

[1] In 2022, Mau, Lux, and Westheuser found in a representative survey of 2,530 people aged 16 and over that 28% of respondents agreed with the statement that gender-inclusive language is important for equality (Mau, Lux & Westheuser, 2023, p. 186). The results of a 2021 survey of 10,062 people by Lempp, Serfling, and Rolf are somewhat more nuanced. To the question, "Should more emphasis be placed on gender-inclusive language?" 2 percent of CDU supporters, 7 percent of SPD supporters, 20% of Greens, 4 percent of FDP supporters, 11% of Left Party supporters, and 1 percent of those close to the AfD *definitely* agreed (Lempp, Serfling & Rolf, 2023, p. 251). I also agree with this question, so I belong to a small minority and feel comfortable with that.

[2] From 1936, the "Reich Central Office for the Fight Against the Gypsy Menace" collected data on Sinti and Roma.

between 220,000 and 500,000 Sinti and Roma fell victim to the genocide planned by the National Socialists (Sciba, 2015).

"Sinti" is the self-designation of members of the minority who have lived in Central Europe and Germany since the late Middle Ages; "Roma" refers to the larger part of that minority who immigrated to Central Europe from Eastern and Southeastern Europe (Kelch, 2018, pp. 32). About 70,000 Sinti and Roma live in Germany. In a study by the working group "RomnoKher," 41.2% identified as "Sinti*zze" and 22.9% as "Rom*nja." 9.2% rejected any self-designation. 4.3% used the exonym "Zigeuner" (Strauß, 2021). However, the majority of Sinti and Roma reject the term "Zigeuner" as discriminatory. That should be reason enough to refrain from using the term "Zigeuner" and also from "Zigeunerschnitzel." Call it "paprika schnitzel" if you like, eat "Wiener schnitzel," or become a vegan.

But that does not solve the problem. Under the term *antigypsyism,* or better: "devaluation of Sinti and Roma," racist prejudices against Sinti and Roma have been researched for several years (e.g., Heitmeyer, 2012; Decker et al., 2022). And it repeatedly becomes apparent: resentments against Sinti and Roma are apparently widespread in both East and West Germany. According to the authors, the differences between East and West are significantly to the disadvantage of respondents from East Germany. How can this be explained? It cannot be due to the physical presence of Sinti and Roma in East Germany. Perhaps it is due to the longevity and rigidity of racist traditions and prejudices. But then this would also have to apply to regions of the former West Germany. It seems more likely that prejudices against Sinti and Roma are among the "anti-modern" resentments with which some people in East Germany attempt to express their hostility to democracy and their dissatisfaction with democratic institutions in Germany. In the studies by Decker and colleagues, antigypsyist prejudices correlate significantly negatively, albeit at a relatively low level, with, among other things, "satisfaction with democracy," "trust in the Bundestag," or the "federal government" (Decker et al., 2022, p. 197).

The case of the "Mohrenapotheke" (Moor's Pharmacy) is somewhat different from that of the "Zigeunerschnitzel." There are or were around 100 pharmacies in Germany named "Mohrenapotheke," in Thuringia (e.g., in Erfurt, Mühlhausen, or Rudolstadt), in Bavaria (e.g., in Munich, Coburg, or Bamberg), in Schleswig-Holstein (e.g., in Kiel), or in North Rhine-Westphalia (e.g., in Gütersloh). Sometimes, not far from pharmacies, you can also find "Mohrenstraße" (Moor Street), a "Café zum Mohren" (Café at the Moor), or the "Hotel zum Mohrenbrunnen" (Hotel at the Moor's Fountain), where, for example, you can drink an "Eisenberger Mohrentropfen" (Eisenberg Moor's Drop). Karl Marx was called "Mohr" by his children and friends because of his dark eyes and initially dark hair. "Dear Mohr" was the salutation with which Friedrich Engels often began his letters to Karl Marx (e.g., Engels, 1974, p. 9; original: 1860). People socialized in the GDR may still remember the children's book and the film of the same name "Mohr und die Raben von London" ("Mohr and the Ravens of London").

But the "Mohr" has done its duty, in several respects, and certainly as a designation. At the latest since the *Black Lives Matter movement* reached Europe and Germany in 2020, anti-racist discourses have also revolved around the naming of businesses, products, street names, and monuments.

"Mohr" (for example, "Mohrenstraße" in Berlin, which is about to be renamed), "Nettelbeckufer" (a street in Erfurt), Nettelbeckplatz in Berlin, Nettelbeckstraße in Dortmund, "Neger" (a term introduced into the German language in the 17th century and *always* meant as a discriminatory term for people with dark skin), and numerous other exonyms are either rejected by anti-racist scholars, politicians, or activists as signs of racist discrimination, with calls for renaming, or such terms are vigorously and with historical sources defended as appropriate (cf. Herling, 2021).[3] For example, the Nettelbeckufer.

In 1807, Joachim Nettelbeck played an important role as a citizen adjutant in the defense of the Prussian fortress of Kolberg against Napoleon's troops. He was also a helmsman on a Dutch slave ship, traded slaves himself, and was instrumentalized as a great hero by the Nazis in a UFA propaganda film. In this film, premiered on January 30, 1945, Heinrich George played Nettelbeck under the direction of Veit Harlan (Kemmerer & Nettelbeck, 2021). Nettelbeck cannot be held responsible for this instrumentalization. However, Nettelbeck's "dark" side as a slave trader prompted initiatives such as "Decolonize Erfurt" to vehemently demand the renaming of streets or squares named after Nettelbeck (see e.g. Berlin.de, 2023; Deutschlandfunk, 2022). On September 6, 2023, Erfurt's Nettelbeckufer was renamed Gert-Schramm-Ufer.[4] However, such initiatives are not without a certain narrow-mindedness (see *Chap. 13*). As long as, in places where people live on Nettelbeckufer or Nettelbeck-Platz, for example, Bismarck

---

[3] In the aforementioned study by Mau, Lux, and Westheuser, 87% of respondents agreed with the statement: "Renaming streets because of racism is exaggerated" (Mau, Lux & Westheuser, 2023, p. 186).

[4] Gert Schramm (1928–2016) was the youngest of a total of six Black prisoners incarcerated in the Buchenwald concentration camp.

monuments (such as at Anger in Erfurt[5] or in Berlin's Tiergarten) can still be "admired," these renaming initiatives simply lack consistency.

By 1880 at the latest, the Germans were trying to secure a colonial "place in the sun," in Africa or the Pacific, usually in concert with the other European colonial powers. At the so-called "West Africa Conference," to which Otto von Bismarck invited representatives from thirteen European states, the USA, and the Ottoman Empire to Berlin from November 15, 1884, to February 26, 1885, the colonial division of Africa, or rather the world, was effectively decided in a final act. Otto von Bismarck declared the territory of present-day Namibia a "protectorate of German South West Africa," that is, a German colony. The indigenous Herero and Nama were downgraded and disenfranchised by the colonizers to second-class people. They lost their land and thus the most important basis of their existence. In January 1904, the Herero rose up and resisted oppression. Under the command of Lieutenant General Lothar von Trotha (1848–1920), the Imperial Protection Forces brutally crushed the uprising. The survivors were interned by von Trotha in concentration camps or driven into the desert, where most died of thirst. In October 1904, the Nama tribe also rose up against the German colonial rulers. Their struggle also ended in defeat in 1907. Approximately 80% of the Herero and more than half of the Nama lost their lives in these colonial wars. It was genocide—the deliberate, inhumane, indeed inhuman extermination of ethnic groups. Bismarck had already died by this time. Nevertheless, he counts among the activists of colonization. In Germany, there are said to be around 700 Bismarck towers, columns, and other monuments to Bismarck (Wolfrum, 2002, p. 4).

To not lose sight of the "Mohrenapotheke" and "Mohrenstraße." In recent years, numerous Berlin initiatives have vehemently opposed the latter designation. The Berlin CDU opposed a renaming, as did the historian Götz Aly. He sued the Berlin Mitte district office and lost (Aly 2021; Berlin Administrative Court, 2022).

[5] The larger-than-life bronze statue of the "Iron Chancellor" at Erfurt's Anger was erected in 1904, removed in 1948, and re-erected in 2004 at the initiative of a Bismarck Tower Association (Erfurt-Web, 2023).

Now, "Mohrenstraße" is to be renamed after the first known philosopher of African descent in Germany, Anton Wilhelm Amo (1703–1759)[6], who, however, had little connection to Berlin.

The "Mohrenapotheke" is a popular subject of dispute, in feuilletons, talk shows, or academic treatises (e.g. Herling, 2021; Munske, 2020). Sandra Herling (2021) took the trouble to analyze the lay-linguistic "Mohr discourses" in internet forums between June 2020 and June 2021. The results, in brief: Some demand a renaming because the term "Mohr" is outdated and racist; others insist on continuing to call it "Mohrenapotheke," arguing that the reference to the "Mohr" or "Moors" signifies appreciation for healers from the "Orient."

Years ago, Peter J. Bräunlein (1991) attempted to find more or less plausible answers to the question of the naming of "Mohrenapotheken." He examined two main possibilities in detail: The name "Mohr" could refer to the captain Mauritius, who came from Upper Egypt and served under the Roman Emperor Diocletian (between 236 and 245 to 312). According to legend, this soldier, now venerated as a saint, commanded a unit of Christian soldiers who, together with him, supposedly died a "hero's death" in a "cleansing action" ordered by Diocletian. Later, Mauritius became the patron saint of soldiers, weapons and knife makers, dyers, glass painters, grocers, etc. (Bräunlein, 1991, pp. 224). On city coats of arms (e.g., Coburg), in the coat of arms of the Nuremberg Tucher family, or as a saint's figure in Magdeburg Cathedral, Mauritius is depicted as a person with dark skin and curly hair. However, there is no historical evidence that he was or is a classic patron saint of pharmacies. Somewhat more plausible, though still on shaky ground, is Bräunlein's attempt to establish a connection to the "Three Wise Men." Of the three who are said to have brought gold, frankincense, and myrrh to the Christ child, Caspar is usually depicted as dark-skinned. It could have been this very Caspar who brought myrrh, a medicinal substance, as a gift. Thus, it is possible that in the 16th century an associative chain developed in which the "Mohr" Caspar and his myrrh became the basis for the connection between doctor, Moor, and pharmacy. Not bad.

The German word "Mohr" is derived from the Latin "maurus" and the Greek "mauros." Both mean "black," "dark." As a designation, the word "Mohr" initially referred to the inhabitants of Ethiopia and later to people from western

[6]Amo studied in Halle and Wittenberg, earned his doctorate there in 1734 on the mind-body problem (De humanae mentis apatheia), and from 1736 taught in Halle and in 1739 in Jena as a lecturer in philosophy (Martin Luther University Halle Wittenberg 2023).

North Africa, from Mauretania. From there, according to Peter J. Bräunlein (ibid., p. 219), the non-Christian conquerors of the Iberian Peninsula also came in the 8th century AD, that is, the enemies of the Christians. It is therefore not surprising that over the centuries the term "Mohr" not only became a term for people with dark skin, but also developed into a derogatory term for people of color. Not even the "noble Moors" or the "court Moors," who were kept by nobles and the wealthy at European courts, could prevent this (Kunz, 2021, p. 57). In this respect, the "Mohrenapotheke," the "Mohrenstraße," or the "Hotel zum Mohren," whether referring to one, two, or three "Moors," are not only somewhat problematic; they also cannot be viewed independently of the hegemony of Christianity and colonial history. I will return to this point.

We do not want to forget the "Indians" either: Franz Kafka had the wish to become an Indian: "If only one were an Indian, always ready, and on the galloping horse, slanting in the air, again and again briefly trembling above the trembling ground…" (Kafka, 2010, p. 17; original: 1912)[7]. There is no need to recall the error of Christopher Columbus. The everyday discrimination against the indigenous peoples of the Americas is also well known, as is their ongoing struggle for their land and their lives. When people speak of "Indians," in most cases there is no racism implied, even if one might question whether the exonym is particularly appropriate. At a Green Party convention in March 2021, Bettina Jarasch, a member of the Berlin Senate, was asked what she would have liked to become as a child. Her answer was "Indian chief." And a shitstorm broke out. Indigenous tribes, critics said, would find the term "Indian" offensive (Berliner Zeitung, 2021a). Bettina Jarasch did apologize, but the Berliner Zeitung (BZ) persisted and asked politicians, celebrities, and scholars for their views. Heike Bungert, a professor in Münster, was also asked and quoted by the BZ as follows: "As an exonym, the word 'Indian' is now considered colonialist by many Germans. However, the problem of a missing collective term remains. In the USA, 'Indians' tend to refer to themselves as 'American Indians.' In line with this preference, one can still use 'Indian' in German—although the word 'Indigenous' or the specific name of the individual group is to be preferred" (Berliner Zeitung, 2021b).

I am not a member of one of the 400 different clubs that meet on weekends in Indian clothing, sleep in tipis, and call themselves redskins (Alvarez & Kunze,

---

[7] I owe the reference to Franz Kafka's fragment as well as further suggestions to the book "Ethics of Appropriation" by Jens Balzer (2023).

2018, p. 83). But I admit that as a child I preferred playing Indian to playing cowboy, and to this day I find the Indian films with Gojko Mitić better than the Karl May adaptations. And my grandchildren read with great interest and much sympathy my now rather worn-out book "Blauvogel—Wahlsohn der Irokesen" ("Bluebird—Adopted Son of the Iroquois") by Anna Jürgen; they watch Yakari's adventures on television or enjoy celebrating Indian-themed birthdays.

"You see—I don't just want to replace one kind of madman with another kind of madman [...] I want to put an end to all delusions and to the human tendencies that support delusions and make it easy for their prophets to succeed" (Feyerabend, 1992, p. 55).

15.3 Exclusion Procedures and Discourse Controls

And then there are the debates about *blackfacing* and *jewfacing*. Blackfacing occurs when white people dress up, painting their faces with dark makeup to convey a stereotypical image of Blackness or of people of color. Against this background, in recent years there has been heated debate in the arts sections and in academia (e.g., Dillmann, 2020; Skwirblies, 2022) about, among other things, the black makeup that white actors apply to their faces in order to play Othello. Similarly heated were the discussions in summer 2023 when it became known that the American, non-Jewish actor Bradley Cooper had modeled a large prosthetic nose onto his face in order to play the conductor and composer Leonard Bernstein in the film "Maestro" (e.g., Tagesschau, 2023). *Jewfacing* was the accusation leveled at the actor on social media, meaning the stereotypical portrayal of Jews by a non-Jew.

The origin of *blackfacing* is seen in the 19th-century American *minstrel shows,* "[...] in which stereotypical imaginations of Blackness amused a white audience and legitimized the exploitation and enslavement of Black people" (Wilmot, 2020, p. 117). And *jewfacing* is known from the propaganda films of the National Socialists, for example from the antisemitic film "Jud Süß" about the Jewish financial advisor to Duke Karl Alexander of Württemberg, Joseph Süß Oppenheimer, played by the non-Jewish actor Ferdinand Marian (for a detailed discussion of racism in the film "Jud Süß" see also: Malli, 2022).

Thus, in both cases—blackfacing and jewfacing—we are dealing, in terms of their historical roots, with forms of racist bodily staging. However, I wonder whether these historical roots always need to be invoked whenever, in theater or film, white non-Jewish people attempt to play the roles of Jews or People of Color. Helen Mirren, a non-Jew, was brilliant in the role of Golda Meir in the 2023 film “Golda”, and Martin Reik was outstanding as a black-painted Othello in “Othello. Venedigsneger” (Deutschlandfunk, 2012).

Admittedly, these are not scientific arguments, but rather the very personal opinions of a scholar who also considers Patti Smith’s “Rock N Roll Nigger” to be a great song. Spotify, Apple Music, Tidal, and Amazon Music have removed this song from their streaming offerings. That, too, is absurd. When Patti Smith sings “Nigger,” and refers to Jimi Hendrix, Jesus Christ, her grandmother, or Jackson Pollock as “nigga,” it is actually clear what is meant and how. It is not about mocking African Americans. Patti Smith virtually deconstructs the word “Nigger” by using it as a symbol for those who are perceived as “outside of society.” “Outside of society. They're waitin’ for me. Outside of society. If you're looking, that’s where you'll find me” (see also: Büttner, 2022).

Let’s go a step further: Do you know the poem “avenidas” by Eugen Gomringer? The poem is written in Spanish and for a long time adorned the façade of the *Alice Salomon University of Applied Sciences* in Berlin-Hellersdorf. The English translation reads: “Avenues/Avenues and flowers/Flowers/Flowers and women/Avenues/Avenues and women/Avenues and flowers and women and/an admirer.” In 2017 and 2018, there was heated debate about whether this poem should remain on the university’s façade or, as the university’s student council demanded, be painted over because it conveyed an “old-fashioned image of women.” It was eventually painted over and can now be found—not far from the *Alice Salomon University of Applied Sciences*—on the façade of a residential building (see also: Süddeutsche Zeitung, 2019). From time to time, this poem is also seen as an indicator of a phenomenon that is called “cancel culture” not only in the arts sections but also in academic publications. “Cancel culture” (from the English “to cancel,” “to call off,” “to drop,” “to delete”) refers to a phenomenon in which “[…] unwelcome, more or less well-known, living or deceased persons (for example from academia, the arts, and politics) are deprived of support or targeted, with the aim of damaging their reputation, preventing them from practicing their profession or the reception of their work, or reducing their presence in mass media and social media” (Bendel, Lin-Hi & Suchanek, 2022, p. 15).

Adrian Daub (2022) analyzes the developments of cancel culture (CC) from its precursors during the Reagan administration, through alleged canceling at American universities, to cancel-culture accusations in far-right and right-wing populist circles. The discourse about CC, according to Daub, is increasingly revealing itself as a revival of the debates about political correctness that gained momentum in the USA in the early 1990s and were rhetorically fueled by both supporters of the Bush administration and Democrats around *Clinton.* Talking about CC seeks and channels attention and relies on anecdotes. These are stories that play with space and time, often imprecise but told as if credible. "Consuming cancel-culture anecdotes means a kind of education in selective tolerance for exaggeration. Or perhaps better, cancel-culture anecdotes suggest which kind of exaggeration is legitimate and which is not" (Daub, ibid., pp. 214). The birthplace of CC is the internet. Only there does the term have a hint of validity.

Well, that may be too much praise or criticism of the internet. Cancel culture has a much longer history than the internet itself. Consider, for example, the "Index Romanus," the "List of Prohibited Books" established by the Roman Inquisition (first published in 1559, last updated in 1962, and then discontinued). The list included, for instance, books by Honoré de Balzac, Giordano Bruno, Auguste Comte, René Descartes, Alexandre Dumas, Gustave Flaubert, Heinrich Heine, Victor Hugo, Immanuel Kant, John Stewart Mill, Jean-Jacques Rousseau, Voltaire, Thomas Hobbes, Moses Maimonides, or Simone de Beauvoir and Jean-Paul Sartre (Wolf & Arning, 2010, p. 165). One might also recall the Roman verdict of "condemnation of dissent" (damnatio memoriae) and its application in the Middle Ages (Schwedler, 2020), or the book burnings at the Wartburg festival of the student fraternities in 1817 (Ries, 2019), or in 1933 throughout Germany (Schenck, 2023). If these examples do not fit the scholarly understanding of cancel culture, let me remind you of the "Kahlschlagplenum," the 11th Plenum of the Central Committee of the SED, where in December 1965 the fight against "enemies of socialism" was decided. Among the enemies, as could be heard from Walter Ulbricht, Erich Honecker, Horst Schumann, and other men at this plenum, were "beatniks," "losers in jeans," "rowdies," "beat groups," but also Wolf Biermann, Werner Bräunig, Robert Havemann, Stefan Heym, and all those who were not willing to create an "[…] art pleasing to the party" (Günter Mittag, quoted in Groschopp, 2012, p. 423). As a result, books, films, plays, music groups, writers, and musicians were banned. Incidentally, a few months before the 11th Plenum, in October 1965, the East German rock group "Butlers" and numerous other bands were banned from performing, which led to a "beat uprising" in Leipzig (Mählert & Stephan, 1996, p. 166). Several thousand young people demonstrated for the reinstatement of the banned bands. And then things took their course. The demonstration was brutally ended by the police and other security forces. For the sake of balance, one should also mention the West German "Radical Decree" of 1972, as a result of which "enemies of the constitution" were no longer allowed to work in the public sector (e.g., in universities, the federal railway, or schools). In total, approximately 3.5 million people were screened, more than 1,200 people were classified as left-wing extremists, and around 260 people were dismissed (see also: Braunthal, 1993).

Let us follow Adrian Daub further: While in the USA there is a shift of cancel culture debates from digital media to print media, in Germany the discourse takes place both in social media and in newspapers—especially in the arts and culture sections. Cancel culture texts say: "It is acceptable to always view individual cases in the light of a very specific framing and then convince oneself that one is informed. In such texts, mood comes before detail; everything is analyzed in the Caravaggesque spotlight of the culture war" (Daub, 2022, p. 281). The interest in alleged censorship, identity politics, and "wokeness" is disproportionate to the demonstrable prevalence of CC. In Germany, cancel culture panic is linked to the fear of "left-wing censorship" and "identity politics from the left." Instead of engaging with the roots of (American) identity politics, with the struggles of African American women or LGBTQ+ people, members of the white majority society accuse those for whom their identity is essential of lacking tolerance, unwillingness to debate, etc. "The criticism of identity politics is in many cases simply identity politics for people named Jürgen" (Daub, ibid., p. 257). The fear of CC always carries with it a certain degree of anti-Americanism. "The fight against cancel culture may see itself as the spearhead of a resilient liberalism. In truth, it is part of the backlash that threatens liberal democracy in the first place" (ibid., p. 341).

As is well known, one does not have to take speeches by German politicians on Ash Wednesday too seriously. Nevertheless, the speech by Bavarian Minister President *Markus Söder* on February 22, 2023, was interesting. "A compulsive use of gender-neutral language," said *Söder*, with an unmistakable glance at what was happening in the German capital and at his main political opponent in Bavaria, the Greens, "[…] we do not do that in Bavaria. In Bavaria, everyone is allowed to be happy in their own way. Here, we do not submit to any re-education fantasies, nor do we practice cancel culture" (Hamburger Abendblatt, 2023).

Unfortunately, *Söder* did not say directly who exactly is responsible for the "compulsive" gendering, the "re-education fantasies," and "cancel culture." However, his remarks illustrate some of the main findings from Adrian Daub's book: There is more whispering about CC than concrete discussion or writing. The word "cancel culture" appears more often among those who feel threatened by events than among those who might actually be canceling.

Another finding is the anecdotal nature that characterizes the debates about CC. An anecdote, as is well known, is a short story about events that actually happened or were invented. The more often an anecdote is told, the more believable it becomes, even if, in the course of retelling, some features of the narrated event change, disappear, and/or are replaced by new features. This also happens in the debates about CC. However, the anecdotes about CC are not as indeterminate in space and time as Adrian Daub suggests. The examples he places at the beginning of each chapter of his book seem to confirm this. In most examples, people are named who were apparently canceled; the locations of the supposed "cancel" events are also rarely left unmentioned. This is usually also true of the examples or anecdotes told about canceling in German contexts (e.g., the white musician *Ronja Maltzahn,* who was declared "unwelcome" at an event in Hanover in spring 2022 because of dreadlocks; the cabaret artist *Lisa Eckhart* was disinvited from the *Harbourfront Literature Festival* in Hamburg in 2020 due to allegedly racist and anti-Semitic remarks; *Bernd Lucke,* co-founder of the AfD, was prevented by students from giving his lecture in October 2019). Regardless of whether these exemplary cases involved CC or were merely disputes, the apparent concreteness in the anecdotes increases their credibility on the one hand. In this sense, CC anecdotes differ from the modern legends and urban myths that Bengt af Klintberg (1990) so aptly deconstructed (see also: Frindte & Frindte, 2022, pp. 75). On the other hand, cancel culture is not a cultural technique practiced by woke, left-wing, and identity-concerned people. Rather, talking and writing about cancel culture mostly belongs to the language games of certain conservative and (right-)populist subcultures.

However, those concerned with identity are by no means free from such language games and instruments of exclusion. Empirical studies show that the effort to exclude individuals or social groups because they allegedly hold offensive, misogynistic, racist, or anti-Semitic positions is not just a rhetorical myth, but can also be influenced by the dominant cultures existing in groups, societies, or nations. Pippa Norris (2023) demonstrates this with a study involving 2,500 scholars from various countries. While in countries such as the USA, Great Britain, or Sweden, it was mainly conservative scholars who faced a "cooler" and more hostile climate, in countries with a more "traditional moral culture," such as

Nigeria, it was especially left-leaning individuals who reported threats of social exclusion. In this context, it should also be remembered that scholars who provided their expertise to the public and policy advice during the COVID-19 pandemic were confronted with accusations of data manipulation, subjected to vile insults, or even threatened with violence.

Against the backdrop of such individual cases, Mitchell G. Ash draws attention to "discourse control at German universities" (Ash 2022). "Discourse control" is, so to speak, the neutral term for cancel culture or political correctness. Discourse control prevails when the—usually highly valued—freedom of science is threatened.

Such control can take place *externally,* for example by institutions outside the space in which research is conducted (I previously referred to this space as the *Context of Manufacture*; *Ch. 14*). Mitchell G. Ash cites as an example a scholarly article by Stefan Lewandowsky (2019), which demonstrates that opponents of anthropogenic climate change behave like adherents of conspiracy theories. The article had already been published in a scientific journal but had to be retracted due to several lawsuits.

Discourse control can also be exercised *internally.* This is the case, for example, when "[…] a discipline or group of disciplines attempts to control the discourse in other disciplines regarding who is allowed to speak about sex and gender, in which fields, and in what manner" (Ash, 2022, p. 22).

Finally, *internalized* forms of discourse control are also conceivable. These function almost like anticipatory obedience. Scientists control themselves by refraining from undesirable research topics, practices, or rules. "Internalization of discourse controls may also occur when external political demands or value systems are elevated to the foundation of a scientific discipline, for example through the dogmatization of gender, diversity, or 'decolonization' perspectives beyond personnel policy, and the scientists involved make these requirements the basis of their own research" (Ash, 2022, p. 28). *Ash* and his colleagues therefore recommend, among other things, not to confuse science with political irresponsibility, to enable scientific discourse without moralization or political taboos, and to take seriously the duty of care regarding personalized campaigns and threats against university members.

To recall Paul Feyerabend:

"Which values do we choose to examine the usefulness of today's sciences? It seems to me that happiness and the full development of individual human beings must still be regarded as the highest value today. This value does not exclude other values that follow from institutionalized forms of life (examples are truth, courage, self-denial, and so on). It can also promote these other values, but only to the extent that they contribute to the progress of the individual. What is excluded is the abuse of institutionalized values to condemn and perhaps even eliminate those people who wish to organize their lives differently. [...] The fundamental value of human happiness and self-perfection therefore demands a methodology and a set of institutions that enable us to lose as little as possible of our abilities and to realize our own inclinations as far as possible" (Feyerabend, 1978, p. 167).

15.4 On Whiteness and Cultural Appropriation

This brings us to the representatives of *Postcolonial Studies, Critical Race Studies,* or *Whiteness Studies,* who, incidentally, regard Immanuel Kant as a theoretical pioneer of imperial conquest, colonialism, and racism (e.g., Dhawan, 2017; Moses, 2021a; Robinson, 2002, and many others).

Among other things, they refer to Kant's racist remarks, which he made in lectures entitled "Physical Geography" in 1772/73 in Königsberg. According to Karl Vorländer (1993, Part 2, p. 65), it must have been a popular and well-attended lecture. However, Kant never published this lecture himself. The text available today is based on lecture notes, which Kant's student and colleague Friedrich Theodor Rink published as a book in 1802—while Kant was still alive. There one reads, among other things: "Humanity is at its greatest perfection in the race of the whites. The yellow Indians already have a lesser talent. The Negroes are much lower, and lowest of all is a part of the American peoples. [...] The inhabitant of the temperate zone, especially its central part, is more beautiful in body, more industrious, more humorous, more moderate in his passions, more intelligent than any other kind of human in the world. Therefore, these peoples have at all times instructed the others and conquered them by force of arms" (Kant, AA, Vol. IX, p. 317).

The heterogeneous, multidisciplinary research fields of Postcolonial or Critical Race Studies are concerned not only with the effects and legacies of colonialism, but trace almost all social, political, and cultural problems of the present, including anthropogenic climate change, back to the colonial dominance of the "white West." The racism of whites is said to be both the cause *and* the consequence of colonial crimes in the past and present. Some representatives of postcolonialism also see antisemitism and the Holocaust as part of colonial history, just like the mass enslavement, displacement, and extermination of colonized peoples. The point is not to deny the Holocaust, but—after the "postcolonial turn"—to deny its singularity and to situate antisemitism and the mass extermination of Jews within the history of colonial racism (Cheyette, 2018). Especially those scholars who study the genocides of indigenous peoples argue that the thesis of the Holocaust's uniqueness is dogmatic, negates the perspectives of the victims of colonial genocides, and promotes a hegemonic Eurocentrism (Moses, 2002).

In Germany, the Australian political scientist A. Dirk Moses, who teaches in the USA, caused a stir in the summer of 2021 when he criticized the resolution of the German Bundestag from May 2019, in which the BDS boycott calls[8] against Israel were declared antisemitic. This resolution was passed with the votes of the CDU/CSU, SPD, FDP, and large parts of the Greens as well as an independent member of parliament (Bundestag, 2019). A. Dirk Moses called this resolution the "most ominous signal to date" of the ignorance with which the German elites attempt to deny the Palestinians' struggle against colonization by Israel (Moses, 2021b).

"For many, the memory of the Holocaust as a rupture of civilization is the moral foundation of the Federal Republic. To compare it with other genocides is therefore considered heresy, an apostasy from the true faith. It is time to abandon this catechism" (Moses, 2021b). According to *Moses,* German scientific elites, together with leading politicians, insist on the uniqueness of the Holocaust and regard it as a "sacred trauma," "[...] which must not under any circumstances be contaminated by other events—such as non-Jewish victims or other genocides—since this would impair its sacred redemptive function."

There is no question that coming to terms with the colonial crimes of the "West" is urgently necessary, that the intellectual pioneers must be named, and that old and new forms of racism must be combated. However, the history of hostility toward Jews begins earlier than that of colonial crimes. And the Holocaust was more than a terrible excess of colonialism.

---

[8] BDS stands for "Boycott, Divestment and Sanctions." Its supporters call for the State of Israel to be boycotted economically, politically, and culturally because of the "occupation and colonization of the Arab land occupied in 1967" (BDS, 2021). It is a transnational campaign that refers to UN resolutions and demands respect for human rights, but ultimately seeks to isolate the State of Israel from the rest of the world. The modern form of colonial racism—according to, for example, Omar Barghouti (2021), a co-founder of BDS—manifests itself in the oppression of Palestinians by Israelis.

Antisemitism is the calculated staging of the annihilation of Jews as Jews, because there is an audience that first marvels at the staging, then willingly participates in it, and finally carries out the annihilation of Jews as Jews itself, and in even more brutal ways. Ultimately, it was always the audience, as populus, as people or as the "common people," who put into action the staged hostility toward Jews (even before the Middle Ages) or the ideologized antisemitism. Thus, "common" Persians were to carry out the annihilation orchestrated by the high priest Haman (Book of Esther); "common" citizens were incited by so-called Crusaders to the pogroms of the High Middle Ages; "common" citizens also killed Jews who were accused of poisoning wells and murdering Christian children; "normal" German students burned Jewish writings at the Wartburg Festival in 1817; perfectly "normal Germans" allowed themselves to be "organized" by the NSDAP and the SA on the night of November 9 to 10, 1938, to destroy Jewish businesses, private homes, apartments, and synagogues; and perfectly "banal" Germans ultimately organized and carried out the Holocaust as well. And as with every bad production, in the end no one in the audience knew anything or had always rejected evil from the start. Lest we forget: two thirds of the Jews living in Europe fell victim to the Holocaust, six million people! And today, it is perfectly "common" people who deny or relativize the crimes of National Socialism, equate Israeli policy toward Palestinians with the annihilation of Jews under National Socialism, call Israeli citizens Nazis, or threaten people who are recognizable as Jews.

Dan Diner, Saul Friedländer, Norbert Frei, Jürgen Habermas, and Sybille Steinbacher vehemently rejected the theses of A. Dirk Moses (Friedländer, Frei, Steinbacher & Diner, 2022). Habermas, among others, points out that comparing the Holocaust with colonial genocides ignores a "specific difference." While the German crimes in the "East" were about the ruthless "acquisition of living space," as a result of which the Slavic peoples living there were oppressed, exploited, and also killed, the deported Jews were murdered "for the sole reason that they were Jews" (Habermas in: Friedländer et al., 2022, p. 11). The annihilation of the Jews was not directed against foreigners, but against their own citizens, "who first had to be identified as a subversive threat and gradually excluded from their own population before they were transported to the extermination camps." This is the difference that makes the difference, as Saul Friedländer also points out. "The difference," says Friedländer, "lies in the historical context of each genocide. In this sense—and only in this sense—the Holocaust is special and truly unprecedented" (Friedländer et al., ibid., p. 18). This context and its historical development—a two-thousand-year-old hatred of Jews—is crucial to understanding the uniqueness of the Shoah. The culture of remembrance related to this does not stem from any kind of "catechism," but developed in Germany very slowly over generations and has nothing to do with a hidden political agenda. Sybille Steinbacher counters Moses by saying that, compared to other genocides, such as those against the Herero and Nama, the Armenians, or the Tutsi, the annihilation of the Jews was unique. It was unique because the "Final Solution to the Jewish Question" did not stand in the tradition of a colonialist program, but was driven solely by its own fanatical lack of restraint (Steinbacher in: Friedländer et al., 2022, p. 64).

Historical comparisons can be helpful, but they can also create blind spots and lead to the trivialization of monstrosity. So what is the purpose of the cultural-revolutionary furor with which Moses and others cry out about the supposed prohibition on comparison? "The attack is directed as much against the hard-won achievements of the culture of remembrance in this country as it is against Israel, which is meant as a stand-in when it comes to the topos of the uniqueness of the Holocaust. Jews must not be a special victim group, because, as it is said, there must be an end to the self-assumed Jewish hegemony" (Steinbacher ibid., p. 66). Quite simply, behind the postcolonialist outpourings of A. Dirk Moses lies an anti-Israeli attitude with an antisemitic slant.

But what about "white people" and *whiteness*? On this, there is not always agreement in the scientific communities—there is more agreement that both *whiteness* and *blackness* are social constructions of race (see, e.g., Tißberger, 2017). On the other hand, *whiteness* always expresses a hierarchical power relationship over all those who are not granted *whiteness*. The systematic abduction and enslavement of people was not only closely linked to the dominant colonial policies of the *whites,* but also to industrialization and the growing prosperity in "white Europe."[9] And white Enlightenment thinkers provided the justifications in a Eurocentric manner. "With Kant's 'emergence of man from his self-imposed immaturity,' not everyone is shown the path to freedom," writes Martina Tißberger. And further: "The civilizational project of modernity does not bring opportunities for development for all, but creates a historyless space, called primitiveness, in which subjects and cultures are located who are excluded

[9] It need not be pointed out here that slavery in Africa or Asia was widespread centuries before colonization by the "West" (see also: Patterson, 2018).

from this civilizational project: the 'primitives' and 'savages'—synonyms for non-whites*—who are simultaneously exploited in the European colonies" (Tißberger, 2017, p. 104).

Whether the European Enlighteners (as is well known, they were not only men[10]) and the values they represented have thus become obsolete, as Axel Honneth fears (Honneth, 2023, p. 46), is not settled among the representatives of *Postcolonial Studies*. The Enlighteners are indeed counted by postcolonial critics among the white men who provided the moral justification for the colonization of the "primitives." Yes, it was white men and women who, in the so-called "West," developed an enlightened humanism centered on the human being, their capacity for self-reflection, freedom, dignity, and equality, as well as humane coexistence. "Postcolonialism, too, remains bound to this small legacy of the Enlightenment, as long as it understands its critique of Western thought as a moral call for cultural reorientation. However much hubris and the worst denial of colonial atrocities may be contained in this philosophical legacy, it must nevertheless remain the pedestal on which we can examine our self-understanding and fundamentally revise it in remembrance of our colonial crimes" (Honneth, 2023, ibid.). Certainly, this enlightened humanism has had and still has its dark sides (in detail: Frindte, 2022).

However, if *whiteness* becomes the "key category of racism" (Schmidt-Linsenhoff, 2004, p. 9), then the emancipatory identity politics of precariously living, marginalized, and discriminated groups itself takes on a racist slant. In this respect, it would be absurd, if not chauvinistic, to forgo Western, Occidental humanism and now rely exclusively on the identity politics of nonwhite groups and communities.

I am an old, white man. If one takes the extreme postcolonial critics at their word, I not only bear the burden of colonialism; all my knowledge, feeling, speech, and action are said to be tainted by this burden. "White identity is inherently racist. There are no white people outside the system of white supremacy," writes, for example, Robin DiAngelo, a (white) American sociologist, in her bestseller "*White Fragility*," here quoted from the German translation (DiAngelo, 2020, p. 203). And if I deny being racist, that is said to be merely the justification of my implicit racist thinking.

[10] Who still knows Anne Conway (1631–1679), philosopher and correspondent of Leibniz, Emilie du Châtelet (1706–1749), philosopher, physicist, and partner of Voltaire, or the feminist Harriet Taylor Mills (1807–1858), wife of John Stuart Mill, and many others (see also: Hagengruber, 1998).

The only real counter-thesis to this would be one with which the proponents of *Postcolonial Studies* would certainly not agree: “Cultural exchange is a well-established practice. For millennia, cultures have exchanged ideas, technological achievements, art forms, luxury goods, foods, deities, and prostitutes” (Feyerabend, 2005, p. 286).

Would this not also provide a basis for viewing the much-criticized and indeed ambivalent *cultural appropriation (cultural appropriation)* in a new, positive light, and not reducing it merely to “blackfacing,” “white rap,” or “dreadlocks”? For the thesis is not so absurd. Cultural appropriation can *on the one hand* be part of postcolonial processes of discrimination. Jens Balzer calls this “bad” or failed appropriation. “Bad appropriation exploits the aesthetic products of marginalized people from the position of a hegemonic majority society…” (Balzer, 2023, p. 54). However, it is simply inconceivable that any culture could exist that did not arise from the appropriation of preceding cultural forms. Good, reflective cultural appropriation, *on the other hand,* means constructing something new from the various cultural influences, in which cultural boundaries are crossed and the power relations in previously limited spaces are critically examined. From this, an ethics of appropriation could also be derived, “[…] an ethics that joyfully embraces the foreign within the familiar—and for which solidarity in diversity is more important than the war of all against all” (Balzer, ibid., p. 82). I am certain that Paul Feyerabend would also welcome such an ethics.

References

Alvarez, A., & Kunze, S. (2018). Indianer, der Holocaust und die Frage des Völkermords in Deutschland und den USA. In V. Benkert (Hrsg.), *Feinde, Freunde, Fremde?* (S. 83–104). Nomos Verlagsgesellschaft.

Aly, G. (2021). Meine letzte Kolumne. https://www.berliner-zeitung.de/politik-gesellschaft/meine-letzte-kolumne-li.167934. Accessed: Aug. 14, 2023.

Ash, G. A. (2022). Diskurskontrolle an deutschen Universitäten – Bedrohung der Wissenschaftsfreiheit? *Schriftenreihe der Berlin-Brandenburgischen Akademie der Wissenschaften.* Berlin: Berlin-Brandenburgische Akademie der Wissenschaften. https://edoc.bbaw.de/files/3738/BBAW_WiD_21_2022.pdf. Zugegriffen:14. Aug. 2023.

Balzer, J. (2023; Original: 2022). *Ethik der Appropriation.* Bundeszentrale für politische Bildung.

Barghouti, O. (2021). BDS: Nonviolent, Globalized Palestinian Resistance to Israel's Settler Colonialism and Apartheid. *Journal of Palestine Studies, 50*(2), 108–125.

BDS (2021). http://bds-kampagne.de/aufruf/deutschlandweiter-bds-aufruf/. Accessed: Aug. 14, 2023.

Bendel, O., Lin-Hi, A., & Suchanek, A. (2022). *110 Keywords Wirtschaftsethik.* Springer Gabler.
Berlin.de (2023). Der Nettelbeckplatz braucht einen neuen Namen! https://mein.berlin.de/projekte/der-nettelbeckplatz-braucht-einen-neuen-namen/. Accessed: Aug. 14, 2023.
Berliner Zeitung (2021a). Bettina Jarasch erntet Kritik für „Indianerhäuptling"-Aussage. https://www.berliner-zeitung.de/news/gruenen-parteitag-jarasch-ernet-kritik-fuer-indianerhaeuptling-aussage-li.147659. Accessed: Aug. 14, 2023.
Berliner Zeitung (2021b). Diskussion nach Jarasch-Entschuldigung! Darf man nicht mehr Indianer sagen?. https://www.bz-berlin.de/berlin/diskussion-nach-jarasch-entschuldigung-darf-man-nicht-mehr-indianer-sagen. Accessed: Aug. 14, 2023.
Braunthal, G. (1993). *Politische Loyalität und öffentlicher Dienst. Der „Radikalenerlass" von 1972 und die Folgen.* Schüren Verlag.
Bräunlein, P. J. (1991). Von Mohren-Apotheken und Mohrenkopf-Wappen. *Zeitschrift für Kultur-Austausch-Regensburg: ConBrio Verl.-Ges., 1962, 41*(2).
Bröder, H. C., Meuleneers, P., & Zacharski, L. (2022). Neue Forschungen zur Genderlinguistik – Genderbewusste Sprache in Diskurs, Grammatik und Kognition: Tagungsbericht zur Auftakttagung des DFG-Projekts „Genderbezogene Praktiken bei Personenreferenzen "am 18. und 19.02. 2022. *Zeitschrift für germanistische Linguistik, 50*(3), 548–556.
Bundestag (2019). Antrag: BDS-Boykottaufruf verurteilen. https://www.bundestag.de/webarchiv/presse/hib/2019_05/643058-643058. Accessed: Aug. 14, 2023.
Büttner, J.-M. (2022). Warum Patti Smith das N-Wort verwendete. https://www.derbund.ch/warum-patti-smith-das-n-wort-verwendete-726419595417. Accessed: Aug. 14, 2023.
Cheyette, B. (2018). Postcolonialism and the study of anti-semitism. *The American Historical Review, 123*(4), 1234–1245.
Cirksena, F., & Leiner, D. J. (2022). Priming von Stereotypen durch geschlechtergerechte Sprache in journalistischen Texten. *Studies in Communication and Media, 11*(2), 240–277.
Daub, A. (2022). *Cancel Culture Transfer. Wie eine moralische Panik die Welt erfasst.* Suhrkamp.
Decker, O., Kies, J. Heller, A., & Brähler, E. (Hrsg.) (2022). *Autoritäre Dynamiken in unsicheren Zeiten.* Psychosozial-Verlag.
Deutschlandfunk (2012). Und alle Welt schaut zu. https://www.deutschlandfunkkultur.de/und-alle-welt-schaut-zu-100.html. Accessed: Aug. 14, 2023.
Deutschlandfunk (2022). „Nettelbecks koloniale Aktivitäten waren total unbekannt". https://www.deutschlandfunkkultur.de/zur-umbenennungs-debatte-der-film-kolberg-und-die-nettelbeck-kontroverse-dlf-kultur-45f02f87-100.html. Accessed: Aug. 14, 2023.
Dhawan, N. (2017). Die Aufklärung retten: Postkoloniale Interventionen. *Zeitschrift für Politische Theorie, 7,* 249–255.
DiAngelo, R. (2020). *Wir müssen über Rassismus sprechen.* Hoffman & Campe.
Dillmann, D. (2020). Warum Blackfacing immer noch rassistisch ist. https://www.fr.de/panorama/warum-blackfacing-immer-noch-rassistisch-10960317.html. Accessed: Aug. 14, 2023.
Engbring-Romang, U. (2014). *Ein unbekanntes Volk? Daten, Fakten und Zahlen Zur Geschichte und Gegenwart der Sinti und Roma in Europa.* Bundeszentrale für politische Bildung.

Engels, F. (1974; Original: 26. Januar 1860). Engels an Marx in London. In Karl Marx & Friedrich Engels, Werke, Band 30. Dietz.

Erfurt-Web (2023). Bismarckdenkmal am Anger. https://erfurt-web.de/Bismarckdenkmal_Anger_Erfurt. Accessed: Aug. 14, 2023.

Feyerabend, P. K. (1958). An attempt at a realistic interpretation of experience. *Proceedings of the Aristotelian Society, 58,* 143–170.

Feyerabend, P. K. (1970). Philosophy of Science: A Subject with a Great Past. In R. H. Stuewer (Hrsg.), *Historical and Philosophical Perspectives of Science. Minnesota Studies in the Philosophy of Science, 5.*, S. 172–183. University of Minnesota Press: Minneapolis.

Feyerabend, P. K. (1978). *Der wissenschaftstheoretische Realismus und die Autorität der Wissenschaften.* Braunschweig/Wiesbaden: Vieweg & Sohn.

Feyerabend, P. K. (1981; Original: 1962). *Probleme des Empirismus.* Braunschweig/Wiesbaden: Vieweg & Sohn.

Feyerabend, P. K. (1987). Putnam on Incommensurability. *The British Journal for the Philosophy of Science, 38*(1), 75–81.

Feyerabend, P. K. (1992). *Über Erkenntnis.* Frankfurt a. M./New York: Campus Verlag.

Feyerabend, P. K. (2005). *Die Vernichtung der Vielfalt. Ein Bericht,* herausgegeben von Peter Engelmann. Passagen Verlag.

Feyerabend, P. K. (2009). *Naturphilosophie*, herausgegeben von H. Heit und E. Oberheim. Suhrkamp.

Friedländer, S., Frei, N., Steinbacher, S., & Diner, D. (2022). *Ein Verbrechen ohne Namen: Anmerkungen zum neuen Streit über den Holocaust.* C.H. Beck.

Frindte, W. (2022). *Quo Vadis, Humanismus?* Springer.

Frindte, W., & Frindte, I. (2022). *Support in times of no support. A social-psychological search for traces.* Springer.

Gauck, J. (2019). Gauck kritisiert gesellschaftlichen Diskurs. https://www.tagesspiegel.de/politik/auch-linksliberale-mussen-toleranz-fur-andersdenke-lernen-6684676.html. Accessed: Aug. 14, 2023.

Groschopp, H. (2012). *Der ganze Mensch. Die DDR und der Humanismus – Ein Beitrag zur deutschen Kulturgeschichte.* Marburg: Tectum Verlag.

Hagengruber, R. (1998). *Klassische philosophische Texte von Frauen.* Deutscher Taschenbuchverlag.

Hamburger Abendblatt (2023). https://www.abendblatt.de/politik/deutschland/article237715807/Soeder-will-beim-Politischen-Aschermittwoch-nicht-Gendern.html. Accessed: Aug. 14, 2023.

Heitmeyer, W. (Hrsg.) 2012. *Deutsche Zustände, Folge 10.* Suhrkamp.

Herling, S. (2021). Laienlinguistische Namenkritik im Kontext der Black Lives Matter-Bewegung. In K. Hengst (Hrsg.), *Namenforschung und Namenberatung* (S. 329–360). Leipzig: Universitätsverlag.

Honneth, A. (2023). Ein Sockel muss bleiben. *Die Zeit,* 7. Juni 2023, S. 46.

Kafka, F. (2010). *Kleine Formen,* gesammelt und gelesen von Fritz Michel und Hartmut Abendschein. edition taberna kritika.

Kant, I. (AA, Band IX). Physische Geographie, Vom Menschen. In Kant, I. *Gesammelte Schriften, Band IX,* S. 311–320, Hrsg.: Bd. 1–22 Preußische Akademie der Wissenschaften, Bd. 23 Deutsche Akademie der Wissenschaften zu Berlin, ab Bd. 24 Akademie der Wissenschaften zu Göttingen. Berlin: Walter de Gruyter & Co.

Kelch, C. G. (2018). Dr. Hermann Arnold und seine »Zigeuner«. Zur Geschichte der „Grundlagenforschung" gegen Sinti und Roma in Deutschland. Dissertation. Erlangen-Nürnberg: Friedrich-Alexander Universität.

Kemmerer, A., & Nettelbeck, J. (2021). Joachim Nettelbeck. Des Seefahrers Widersprüche. *Zeitschrift für Ideengeschichte, XV, 1,* 121–124.

Klemperer, V. (1970; Original: 1947). *LTI. Notizbuch eines Philologen.* Leipzig: Verlag Philipp Reclam Junior.

Klintberg, B. af (1990). *Die Ratte in der Pizza – und andere moderne Sagen und Großstadtmythen.* Butt Verlag.

Kunz, S. (2021). Rassismus in aller Munde. Über die Rolle der Sprache im Ent- und Bestehen rassistischer Praktiken. *Wiener Linguistische Gazette, 88,* 53–60.

Lempp, J., Serfling, O., & Rolf, J. N. (2023). *Parteianhängerschaft in Deutschland: Eine Analyse der Parteien und ihrer Anhängerschaften in Bund und Ländern.* Wiesbaden: Springer VS.

Lewandowsky, S. (2019): In whose hands the future? In J. E. Uscinski (Hrsg.), Conspiracy theories and the people who believe them (S. 149–177). Oxford University Press.

Löhr, R. A. (2021). Gendergerechte Personenbezeichnungen 2.0. Wie nichtbinäre Personen den Genderstern und andere Bezeichnungsvarianten beurteilen. *Muttersprache, 131,* 1, S. 172–182.

Mählert, U. & Stephan, G.-R. (1996). *Blaue Hemden – Rote Fahnen.* Opladen: Leske + Budrich.

Malli, D. (2022). Nationalsozialistische Verschwörungsnarrative: Veit Harlans Spielfilm *Jud Süß.* In D. Newiak & A. Schnitzer (Hrsg.), *Verschwörungsideologien in Filmen und Serien.* Springer.

Martin-Luther-Universität Halle Wittenberg (2023). Anton Wilhelm Amo. https://www.amo.uni-halle.de/. Accessed: Aug. 14, 2023.

Mau, S., Lux, T., & Westheuser, L. (2023). *Triggerpunkte – Konsens und Konflikt in der Gegenwartsgesellschaft.* Berlin: Suhrkamp.

MDR (2023). Lob und Kritik für Verbot von Gendern mit Sonderzeichen an Schulen. https://www.mdr.de/nachrichten/sachsen-anhalt/landespolitik/gendern-verbot-schulen-104.html. Accessed: Aug. 17, 2023.

Milton, S. (1995). Vorstufe zur Vernichtung. http://www.ifz-muenchen.de/heftarchiv/1995_1.pdf. Accessed: Aug. 15, 2023.

Moses, A. D. (2002). Conceptual blockages and definitional dilemmas in the „racial century": Genocides of indigenous peoples and the Holocaust. *Patterns of prejudice, 36*(4), 7–36.

Moses, A. D. (2021a). *The Problems of Genocide: Permanent Security and the Language of Transgression.* Cambridge University Press.

Moses, A. D. (2021b). Der Katechismus der Deutschen. https://geschichtedergegenwart.ch/der-katechismus-der-deutschen/. Accessed: Aug. 14, 2023.

Munske, H. H. (2020). *Unser Deutsch II.* Neue Glossen zum heutigen Wortschatz: FAU University Press.

Nübling, D. (2018). Und ob das Genus mit dem Sexus. *Sprachreport, 34*(3), 44–50.

Patterson, O. (2018). *Slavery and Social Death: A Comparative Study.* Harvard University Press.

Payr, F. (2021). *Von Menschen und Mensch* innen: 20 gute Gründe, mit dem Gendern aufzuhören.* Wiesbaden: Springer.

Pöschko, H., & Prieler, V. (2018). Zur Verständlichkeit und Lesbarkeit von geschlechtergerecht formulierten Schulbuchtexten. *Zeitschrift für Bildungsforschung, 8,* 5–18.

Rat für deutsche Rechtschreibung (2021). Geschlechtergerechte Schreibung: Empfehlungen vom 26.03.2021. https://www.rechtschreibrat.com/geschlechtergerechte-schreibung-empfehlungen-vom-26-03-2021/. Accessed: Aug. 14, 2023.

Ries, K. (2019). Die erste „Demo" in Deutschland. Das Wartburgfest von 1817 als radikaldemokratischer Aufbruch. In M. Fröhlich, O. W. Lembcke & F. Weber-Stein (Hrsg.), *Universitas. Ideen, Individuen und Institutionen in Politik und Wissenschaft* (S. 111–133). Baden-Baden: Nomos Verlag.

Robinson, A. R. (2002). Race, place, and space: Remaking whiteness in the post-reconstruction South. *The Southern Literary Journal, 35*(1), 97–107.

Schenck, J. (2023). *Verbrannte Orte. Nationalsozialistische Bücherverbrennung in Deutschland.* Wien, Berlin: Mandelbaum-Verlag.

Schmidt-Linsenhoff, V. (2004). Weiße Blicke. Bild- und Textlektüren zu Geschlechtermythen des Kolonialismus. In V. Schmidt-Linsenhofl, K. Hölz & H. Uerlings (Hrsg.), *Weiße Blicke. Geschlechtermythen des Kolonialismus* (S. 8–18). Jonas.

Schwedler, G. (2020). *Vergessen, Verändern, Verschweigen. damnatio memoriae im frühen Mittelalter.* Köln: Böhlau.

Schwicker, J. H. (1883). *Die Zigeuner in Ungarn und Siebenbürgen.* Verlag Karl Prochaska.

Sciba, A. (2015). Der Völkermord an Sinti und Roma. Berlin: Deutsches Historisches Museum, Berlin. https://www.dhm.de/lemo/kapitel/der-zweite-weltkrieg/voelkermord/voelkermord-an-sinti-und-roma.html. Accessed: Aug. 15, 2023.

Skwirblies, L. (2022). Koloniale Theatralität: Zur Verwobenheit von deutscher Kolonial- und Theatergeschichte. *Thewis, Online-Zeitschrift der Gesellschaft für Theaterwissenschaft,* S. 102–113. https://www.thewis.de/article/download/114/35. Accessed: Aug. 14, 2023.

Strauß, D. (2021). *RomnoKher-Studie 2021. Ungleiche Teilhabe. Zur Lage der Sinti und Roma in Deutschland.* Mannheim: RomnoKher.

Tagesschau (2023). Bernsteins Kinder verteidigen Bradley Cooper. https://www.tagesschau.de/ausland/amerika/bradleycooper-bernstein-jewfacing-100.html. Accessed: Aug. 14, 2023.

Tißberger, M. (2017). *Critical Whiteness.* Springer VS.

Vorländer, K. (1993; Original: 1924). *Immanuel Kant – Der Mann und das Werk.* Felix Meiner Verlag.

Verwaltungsgericht Berlin (2022). Mohrenstraße: Umbenennung nur von Anwohnern angreifbar (Nr. 32/2022); Accessed: Aug. 14, 2023.

Wilmot, V. M. (2020). Das M-Wort als Ausdruck von Kolonialität – anhand eines aktuellen Beispiels aus Thüringen. *Wissen schafft Demokratie, 7,* 110–122.

Wippermann, W. (1998). *Antiziganismus – Entstehung und Entwicklung der wichtigsten Vorurteile.* Stuttgart: Landeszentrale für politische Bildung.

Wittgenstein, L. (1984). *Philosophische Untersuchungen,* Werkausgabe Bd. 1. Suhrkamp.
Wittgenstein, L. (1988). *Remarks on the Philosophy of Psychology.* Vol. I u. II. Chicago: University of Chicago Press.
Wolf, H., & Arning, H. (2010). Die Münsteraner Forschungen zum „Index der verbotenen Bücher", Eine Zwischenbilanz zum DFG-Langfristvorhaben „Buchzensur durch Römische Inquisition und Indexkongregation in der Neuzeit (1542–1966)". *Jahrbuch für Kommunikationsgeschichte, 12,* 165–185.
Wolfrum, E. (2002). *Geschichte als Waffe: Vom Kaiserreich bis zur Wiedervereinigung.* Vandenhoeck & Ruprecht.

16 Corona—Scandal, Crisis, Catastrophe

Some call a crisis a crisis, for others it is a catastrophe. Still others see in the crisis a war that can become a catastrophe. In light of this, there are more important problems in the world than the crises of psychology or the debate over cancel culture. That is a trivial observation. "The real problems of our time […] consist in war, violence, hunger, disease, and environmental disasters" (Feyerabend, 2005a, p. 295).

16.1 Realities

On March 24, 2020, as the coronavirus crisis was just gaining momentum, the inventor of Asterix and Obelix, Albert Uderzo, died of a heart attack. Three years earlier, in the volume "Asterix in Italy," he had given the villain and adversary of the brave Gauls the name Coronavirus. "He is a cunning charioteer with a golden mask, who is supposed to win a horse-drawn carriage race across Italy on behalf of Julius Caesar. Asterix and Obelix are determined to prevent the victory of this malicious 'Coronavirus' at all costs. One of the drawings in the comic shows a cheering crowd shouting 'Coronavirus, Coronavirus'" (Süddeutsche Zeitung, 2020). In the German edition of the comic book, the name does not appear. There, the Roman villain is called Caligarius (in English: the shoemaker).

In December 2019, Covid-19[1] was first identified in Wuhan Province, China. The first case of infection in Germany was confirmed on January 28, 2020.

[1] I predominantly use the term Covid-19, relying on the name introduced by the World Health Organization (WHO) and the Robert Koch Institute.

W. Frindte, *Against Narrow-Mindedness and Chauvinism—Journeying Through Absurd Times with Paul K. Feyerabend*,
https://doi.org/10.1007/978-3-658-50113-6_16

Between March and May 2020, about 140,000 people in European countries died from or with Covid-19. Hospitals in Italy, as well as those in Spain and France, reached their capacity limits, and doctors and nursing staff reached their physical and psychological limits. By mid-April 2020, almost 17,000 employees of the Italian healthcare system had contracted the coronavirus. According to the Italian Medical Association and Italian media, by mid-April, 125 doctors and more than 30 nurses had died after being infected with Covid-19 (Merkur, 2020). In March and April 2020, the Italian region of Lombardy, especially the province of Bergamo, became the area in Europe hardest hit by the coronavirus pandemic. More than 16,000 people died there from Covid-19. The crematoria in Bergamo were no longer able to bury all the dead. Military vehicles transported the coffins to other cities. Doctors and nursing staff in Italy, France, and Spain were forced to make rapid decisions about which Covid-19 patients had the best chances of survival and therefore could receive a place in intensive care, and who was condemned to die. On March 17, 2020, the heads of government of the EU countries agreed to close the EU's external borders and initially ban non-EU citizens from entering EU countries for 30 days. In Germany, from mid-March 2020, numerous shops, schools, daycare centers, and universities were closed, religious services and club meetings were banned, and playgrounds were closed. The lockdown began.

Over the following months, the infectious disease developed into a pandemic, a worldwide spread that ultimately did not stop at national borders.

While the federal government in Germany imposed stricter "rules on the coronavirus" on March 22, 2020 (The Federal Government, 2020), a dispute was taking place in the supposed scientific communities that, even then, seemed somewhat *absurd* from the perspective of laypeople (among whom I count myself). It was about face masks, gloves, and similar protective equipment. On March 3, 2020, the German Federal Ministry for Economic Affairs and Energy ordered an export ban on medical protective equipment (face masks, gloves, protective suits, etc.). This regulation was lifted again on March 19, as the European Commission then imposed such an export ban for all EU member states (BMWK, 2020). At the same time, it became apparent that face masks, protective gloves, and suitable disinfectants were hardly available anymore. They became scarce goods—even in German clinics and doctors' offices. At the end of March, the then Minister of Health Spahn did not consider a general mask mandate in Germany to be necessary. Virologists and doctors were also divided at the end of March/beginning of April 2020 regarding the wearing of masks in public spaces.

The emergency director of the World Health Organization, Michael Ryan, warned that improper putting on and taking off of masks could pose infection risks. Andreas Gassen, chairman of the National Association of Statutory Health Insurance Physicians, said that a possible mask mandate was merely "pure symbolic politics" (Handelsblatt, 2020). The virologist from Halle, Alexander Kekulé, on the other hand, considered masks to be "absolutely sensible." Karl Lauterbach, already the SPD's health expert at the time, saw such sense only if they were high-quality medical masks. And the virologist Christian Drosten initially, at the end of January, expressed skepticism about the protection provided by masks (Drosten, 2020), only later to recognize even homemade masks as a smart idea. The Robert Koch Institute initially, in February 2020, saw only a psychological effect in wearing masks, but changed its assessment at the beginning of April and recommended that even people without symptoms wear masks to reduce the risk of possible infection (Die Presse, 2020).

There is certainly something to it when Paul Feyerabend claims in another context: "It is absurd first to declare that a society serves the needs of »the people« and then to let autistic experts (liberals, Marxists, Freudians, sociologists of all persuasions) decide what »the people« »really« need and want" (Feyerabend, 2005b, p. 151; original: 1987).

The skepticism regarding the protective function of mouth-and-nose coverings can be interpreted as an expression of a lack of plausibility thinking, but also as an indication of contentious presuppositions in science, politics, and everyday life. With Feyerabend, one could also claim: "Scientists do not study all phenomena, but only those from a precisely defined area; and they do not examine all aspects of the phenomena thus selected, but only those that help them achieve their often very limited goals" (Feyerabend, 1985, in: Feyerabend & Thomas, 1985, p. 326).

That it is plausible and not *absurd* to cover one's mouth and nose under certain circumstances could actually have been known from the times of flu epidemics. Or not? The contentious presuppositions in politics, on the other hand, were probably mainly related to supply shortages. In the spring, there simply were not enough masks to buy. In 2020, politicians took advantage of this, enriching themselves through bribery and corruption in the procurement of masks. Also an *absurd* story (see also: Frankfurter Rundschau, 2021).

With the rapid, life-threatening spread of the virus, the names of the disease, its interpretations and explanations, as well as the social movements that supported or opposed the measures proposed, developed, and implemented at scientific, economic, political, and societal levels, also multiplied. Some of the names, interpretations, and explanations were *absurd, narrow-minded* or *chauvinistic*. More on this in a moment.

(Societal) crises can be understood as indicators of "great transformations" (Polanyi, 2019; original: 1944), through which both spaces of possibility for new patterns of interpretation and value deformations and uncertainties can arise. Within this ambivalence, politicians, scientists, journalists, ordinary people, and social movements compete for interpretive authority or for suitable interpretations, attributions (assignments of responsibility), and for appropriate framings of these crises. Words, names, metaphors, myths, and language games play an outstanding role in these struggles over definition and interpretation. Is it an international risk, a crisis that is likely to result in "major system transformations," or even a catastrophe that affects us all?

Interpretive battles and disputes during the pandemic revolved, among other things, around questions such as: Is the Corona crisis an opportunity? This was the subtitle (with a question mark) of a study conducted as part of the Socio-Economic Panel (SOEP), a representative longitudinal survey of private households in Germany, between April and June 2020. In response to the question posed in the subtitle, the authors conclude, among other things: "Overall, the decidedly optimistic and positive results of the present analysis cannot, of course, obscure the massive life- and existentially threatening consequences of the spread of the coronavirus. Nevertheless, despite the great challenges and individual fates, the crisis also seems to offer an opportunity to sustainably strengthen social cohesion" (Kühne et al., 2020, p. 15).

Or is/was the Corona crisis actually a major scandal? If one thinks of the parties that the then British Prime Minister held at his residence at 10 Downing Street while the country was in lockdown, or of the bribery of members of the Bundestag and state parliaments from the CDU and CSU—the so-called various mask scandals—there is certainly something scandalous about it. Journalist Sven Magnus Hanefeld also saw an even deeper "scandal" revealed by the Corona crisis. Not entirely free from conspiracy narratives, he claims, among other things, that a public discourse on measures against the pandemic did not take place; even the "so-called free press" knows only one opinion (Hanefeld, 2020).

The Corona crisis was also interpreted as a "power struggle." Keywords: "USA versus China," as the title of a "Phoenix Runde" on May 7, 2020; "Drosten versus Streek" or "Bild versus Drosten" (for example, the analysis by BILDblog, 2020, a "watchblog" that critically scrutinizes BILD publications).

The war metaphor was also invoked to stage oneself during the Corona crisis. Then US President Donald Trump, like French President Emmanuel Macron, saw himself "at war with a deadly virus" (Tagesschau, 2020a). Alexander Gauland, from the AfD, spoke in the Bundestag in October 2020 of "war propaganda," with which a kind of "war cabinet" (meaning the federal government) was allegedly trying to enforce measures against the Corona pandemic (Zeit Online, 2020). Federal President Frank-Walter Steinmeier opposed such war rhetoric and stated in a televised address on April 11, 2020, among other things: "No, this pandemic is not a war. Nations are not standing against nations, soldiers not against soldiers. Rather, it is a test of our humanity. It brings out the worst and the best in people. Let us show each other the best in us!" (Tagesschau, 2020b). Paul Widmer, a former diplomat from Switzerland and guest columnist for the *Neue Züricher Zeitung am Sonntag* from 2016 to 2021, rightly pointed out that talk of war in connection with the Corona pandemic is immoral and a mockery of people who truly suffer from violence, destruction, and war (Widmer, 2020).

The coronavirus, SARS-CoV-2 or Covid-19, according to UN Secretary-General António Guterres on March 31, 2020, is "the greatest test we have faced together since the founding of the United Nations" (UN Information Center, 2020).

In March and April 2020, the US Congress and Senate passed massive aid packages to accelerate the development of a vaccine, provide financial relief for employers, offer financial assistance to businesses, and support the healthcare system. In mid-March, Donald Trump declared that the virus "doesn't stand a chance against us." At the beginning of April 2020, after religious institutions, schools, and universities had been closed in some states and—such as in California—lockdowns had been imposed to contain Covid-19, he wanted to reopen his beautiful country with a bang. During a press conference on April 23, 2020, Trump mused about whether one could irradiate the body with ultraviolet or some other strong light to kill the corona viruses. Disinfectants, Trump said, could also kill the virus in a minute. Therefore, it would be interesting if one could find a way to inject disinfectants directly—essentially as an internal cleansing.

Three months later, during a campaign appearance in Phoenix, Arizona, Donald Trump said he knew 19 or 20 names for the virus and, together with his cheering supporters, settled on calling it "Kung Flu" in the future, in order, as a White House spokeswoman later stated, to draw attention to the origin of the virus. At that time, there were officially 2.3 million Covid cases in the USA, and more than 120,000 people had died with or from Corona. In mid-September 2020, Trump expressed his conviction that Covid-19 would go away even without vaccines. There would soon be, Trump said, a "herd mentality" (Huffpost, 2020).

Trump probably meant "herd immunity." "Herd mentality" is a term used to describe group-conforming individual behavior, that is, the adaptation of individual behavior and attitudes to supposed norms or demands of social groups. In modern social psychology, the term is not very common. In economic and financial studies, as well as in reflections on user behavior in social media, it occasionally appears to describe the conformist behavior of stock investors or media users (e.g., Hasler, 2011).

On September 20, 2020, seven months after the outbreak of SARS-CoV-2, Johns Hopkins University (JHU) in Baltimore (in the US state of Maryland) reported that 30.81 million people worldwide had been infected with the SARS-CoV-2 virus and affected by the disease Covid-19; more than 957,600 people had died from the disease. The highest number of known infections worldwide was in the USA. There, the 5-million mark was surpassed between August 9 and 10, 2020. And by September 20, more than 6.7 million people in the USA had been infected and over 199,000 had died from Covid-19 (Johns Hopkins University, 2020). On October 26, 2020, a week before the presidential elections in the USA, Donald Trump tweeted: "Cases up because we TEST, TEST, TEST. A Fake News Media Conspiracy. Many young people who heal very fast. 99.9 %. Corrupt Media conspiracy at all time high. On November 4th, topic will totally change. VOTE!" (Trump, 2020) The number of new infections with SARS-CoV-2 in the USA was rising only because so much testing was being done. The corrupt and misleading media, he claimed, were exploiting the pandemic and spreading conspiracy theories (Redaktionsnetzwerk Deutschland, 2020; see also: Frindte, 2022, Chap. 21). *Isn't that absurd?*

On March 10, 2023, JHU ceased recording Covid-19 cases. In Germany and Europe, the pandemic has since been officially declared over (see also Christian Drosten in an interview with the "Tagesspiegel" from December 26, 2022; Drosten, 2022). If you still visit the JHU website, you will see that by the end of statistical recording, more than 600 million people worldwide had been infected with Covid-19. More than 6.8 million people had died from or with Covid-19. At that time, the highest number of official cases was in the USA, with 103.8 million. Since the beginning of the pandemic, 1.1 million people there had died from or with the virus by March 2023 (Johns Hopkins University, 2023).

16.2 Vaccination and Consequences

Researchers at Imperial College London demonstrate in a modeling study (based on data from 185 countries) that worldwide, several tens of millions of lives were saved by COVID-19 vaccinations. Nevertheless, even more lives could have been saved if vaccines had been distributed more rapidly in many parts of the world (Watson et al., 2022). While in Germany, the Standing Committee on Vaccination recommended a fourth booster shot for people aged 60 and over in autumn 2022, on the African continent only one in ten people had received a single vaccination (cited in Quent et al., 2022, p. 69). This is also due to the fact that, although the World Health Organization (WHO) and numerous countries have called for a globally equitable distribution of vaccines, the EU Commission—presumably for political reasons—and the pharmaceutical industry—for profit-oriented reasons—have so far rejected this (AGENDA, 2030, 2023).

Vaccination is never without problems, including vaccination against COVID-19. Mild pain at the injection site, headaches, fatigue, and even fever can occur as side effects after a COVID-19 vaccination. More serious consequences, such as myocarditis or pericarditis, thrombosis, or autoimmune diseases, are also possible. By April 2023, 64.9 million people in Germany had received at least one COVID-19 vaccination; 76.4% of the population. By that time, 18.4 million had opted not to be vaccinated. The Paul Ehrlich Institute (PEI), responsible for vaccine safety, recorded a total of 192,208,062 COVID-19 vaccinations in Germany between December 27, 2020, and March 31, 2023. During this period, 340,282 suspected cases of side effects and vaccine complications were reported. Across all vaccines, 1.8 reports per 1000 vaccine doses were registered; for serious side effects and complications (e.g., myocarditis, pulmonary embolism, facial paralysis, or thrombosis), the reporting rate was 0.3 per 1000 vaccine doses (Paul-Ehrlich-Institut, 2023).

Overall, according to an assessment by the Robert Koch Institute (RKI) in June 2023 (Robert Koch Institute, 2023), the vaccines from BioNTech, Moderna, and Johnson & Johnson provided very high efficacy against severe COVID-19 illness in infections with the Delta variant (which appeared mainly in 2021). Against the Omicron variant, these vaccines were less effective. Therefore, a booster vaccination is recommended. However, it cannot be ruled out that vaccinated individuals may also contract COVID-19. Nevertheless, vaccination can greatly reduce the risk of severe disease progression and prevent the frequency of infections. Systematic reviews and meta-analyses, according to the RKI, show that COVID-19 vaccination not only protects against severe courses of COVID-

19, but can also mitigate the frequency and severity of long COVID symptoms (e.g., shortness of breath, concentration and memory difficulties, muscle weakness, anxiety, and sleep disorders) after a breakthrough infection. Well, that would be good news, if it were not for another: Scientific studies estimate that 10 to 20% of all people who have been infected with COVID-19 continue to struggle with the aforementioned long COVID symptoms weeks after infection (see also Pink & Welte, 2023).

16.3 A Differentiated Clientele

Neither those who chose to get vaccinated nor those who did not could have known about long-Covid symptoms during the pandemic. *Absurd* as it was, the anxious and vaccine skeptics were *narrow-mindedly* lumped together with those who, for political and conspiracy-mythical reasons, took to the streets and social media against vaccination campaigns and political anti-Corona measures, ranting about a Corona dictatorship. Study results suggest that the former do not necessarily think, feel, and act like the latter. For example, a nationwide representative online survey of 3048 adults aged 14 and older, conducted in autumn 2021 by the *Society for Social Research and Statistical Analysis* (Forsa) on behalf of the Federal Ministry of Health, found that those who had not yet been vaccinated could not be painted with the same brush. Among the respondents were a) the "deniers of existence," who believe the Covid-19 virus does not exist at all, b) the "dictatorship assumers," who fear a "Corona dictatorship," c) the "skeptics," who do not doubt Covid-19 but are very skeptical of government measures against the virus and media reports, and finally d) people "not close to Querdenker," that is, people who have nothing to do with the "Querdenker," reject their views, but also believe that the anti-Corona measures could increase the influence of state control. More than one in ten also feared side effects; 16% rejected mandatory vaccination; 15% did not trust the official information on vaccination; 15% stated they feared vaccine injuries and long-term effects; 12% doubted the effectiveness of the vaccines; 10% assessed the risk as higher than the benefit of vaccination. Respondents could name multiple reasons, so the percentages do not add up to 100% (Forsa, 2021).

A similar differentiation is provided by a study from the Westphalian University of Applied Sciences Gelsenkirchen (Wielga & Enste, 2022). The authors surveyed around 1400 adult Germans via an online panel from October to December 2021. Among other things, respondents were asked to assess COVID protection measures (reducing social contacts, keeping distance, voluntary testing, getting vaccinated) and to state reasons if they were against vaccination. Here, too, multiple answers were possible. The most common reasons included references to the "uncertainty of the vaccines" (65%). 38% of the reasons cited pointed to a "lack of necessity," because of trust in one's own health or having recovered status. 28% of the reasons can be categorized as "misinformation," for example, because of references to excessive side effects. 18% of responses can be attributed to "health reasons" (e.g., "forbidden by doctor") and 15% to "external influence" (e.g., vaccination rejected due to political pressure). Doubts about the "effectiveness of the vaccines" were mentioned by 15%. Under "conspiracy theories," the authors summarized 21% of the reasons, which refer to a state-initiated pandemic and vaccination campaign and describe their own vaccine refusal as a just action against state coercion.

At this point, a methodological caveat is appropriate, recalling the uncertainty principle explained in Chap. 14: As scientists, we know that population surveys have their pitfalls. Frequently, questionnaires are used in such cases, which are usually standardized, i.e., consist of a fixed set of questions or statements. Respondents are supposed to answer these questions or evaluate the statements. Standardized response formats, so-called answer scales, are also available for the answers or evaluations. When respondents answer the questions or rate the statements on these scales, researchers must decide whether the answers or ratings actually reflect the respondents' attitudes or whether the respondents are merely reacting to the restrictive situational conditions of the survey in a socially desirable way. Researchers are thus confronted with a decision dilemma that is hardly resolvable. Unless they forgo all standardization and leave it up to the respondents to express themselves as they wish. But even then, the dilemma is not solved. Researchers must again ask themselves what motives underlie the statements of those who express themselves freely. In other words: opinion or attitude researchers are usually observers of something that is vague and idiosyncratic.

The reasons for speaking out against vaccination seem/seemed to be diverse. The same probably applies to the motives for participating in the various and numerous "Corona protests." Between March 2020 and January 2021, according to an estimate by Eric Neumayer and colleagues (2021, cited in: Hunger et al., 2023), there may have been around 1322 demonstrations in Germany, 1487 in

Spain, and 1168 in France. In many cases, these demonstrations were organized, initiated, or at least instrumentalized and promoted by more or less right-wing extremist individuals (e.g., IT entrepreneur Michael Ballweg, see Soldt, 2020) or movements (e.g., the "Querdenken 711" initiative from Stuttgart, supporters of Pegida, the AfD, or the "Reichsbürger"). In a non-representative online survey in autumn 2020 with 1152 adults from Germany and Switzerland, Nadine Frei, Robert Schäfer, and Oliver Nachtwey, among others, concluded that the motives for participating in the Corona protests may be heterogeneous. "Overall, the study participants are neither particularly xenophobic nor Islamophobic, nor are they socially chauvinistic. The attitudinal characteristics even show anti-authoritarian traits: 64% say that children should not be taught to obey authorities. A large proportion want alternative medicine to be placed on an equal footing with conventional medicine, want to return to nature, and to focus more on holistic and spiritual thinking" (Frei et al., 2021, p. 252). Paul Feyerabend might well have been delighted with this "large proportion."

16.4 Of Authoritarians, Conspiracy Mystics, and Narcissists

Carolin Amlinger and Oliver Nachtwey (2022) have attempted to analyze and explain the reasons that may have motivated people to participate in the Corona protests. They are particularly interested in the Querdenker scene, seeing the Querdenker as the prototype of a social phenomenon they call libertarian authoritarianism. An important starting point for their explanation is the studies on the authoritarian character, published in 1950 under the title "The Authoritarian Personality" (TAP) (Adorno et al., 1950).

Amlinger and Nachtwey still see the typical authoritarians at work, but believe they have identified a new, specifically libertarian-authoritarian social character. Libertarian authoritarians are libertarian because they do not identify with an external authority, but form a loose collection of individuals. And they are authoritarian because they direct their resentment at higher authorities and their anger at subordinate groups (women, transgender people, migrants, Jews). "They are libertarian-authoritarian, therefore, because they no longer see themselves bound by any socially obligatory norms, have shed internalized consideration, and are obsessively focused on a perceived danger" (Amlinger & Nachtwey, 2022, p. 178). The libertarian authoritarians are neither conventionalist nor submissive, but rebellious and sometimes aggressive. In the online survey of more than 1150 people from Switzerland and Germany, Nachtwey and colleagues found, among other things: The majority of the Querdenker surveyed do not trust the government, the parties, or the EU, are critical of the "mainstream" media, believe in conspiracy narratives, but are neither xenophobic nor Islamophobic, instead hold rather anti-authoritarian positions, but also agree with the statement that the influence of Jews on politics is too great (Amlinger & Nachtwey, 2022, pp. 257).

Employees of the Berlin Social Science Center surveyed more than 5000 people between June and November 2020. While almost 70% of respondents did not believe the conspiracy myths (e.g., the "great replacement") of the COVID deniers, a good 30% of respondents expressed much or very much understanding for the anti-Corona demonstrations and for the myths about the virus spread there. More than 60% of the "protest sympathizers" locate themselves in the political center, 12.5% of respondents see themselves at the extreme edge of the ideological spectrum, with the larger portion (7.5%) at the far-right edge (Grande et al., 2021). In the continuation of this study, the researchers were able to draw on 12,815 respondents from a total of 16 survey waves (Hunger et al., 2023). About one in five respondents expressed much or very much sympathy for people demonstrating against government measures; around 60% of them would also participate in demonstrations themselves if they had the opportunity. Political distrust is an important motivation here. Far-right orientations and concerns about restrictions on freedom also explain the sympathy for the demonstrations. Perceived economic threats as a result of government-imposed anti-Corona measures are also likely to be associated with this sympathy.

Moreover, there are—how could it be otherwise—relatively robust findings indicating, on the one hand, that the vaccination rate at the federal state level correlates with the willingness to vote for the AfD (Reuband, 2022; Richter et al., 2021). However, among the large group of vaccination supporters, there is not only unanimous approval of a "vaccine mandate," but also skepticism and reservations (Richter et al., 2022). Oliver Decker and colleagues still see research desiderata regarding the motives of unvaccinated people (Decker et al., 2022, p. 96).

In representative surveys with 2000 people each, Tobias Spöri and Jan Eichhorn were able to show "[…] that in 2021 fewer people in Germany subscribed to Corona conspiracy myths than in the previous year. While in 2020 the proportion of respondents who believed in Corona conspiracy narratives was around 15%, it fell to around 9% in 2021" (Spöri & Eichhorn, 2021, p. 4). But they still exist, the conspiracy storytellers. In a study published by the Friedrich Ebert Foundation in 2021, in which 1500 adult Germans were surveyed, 22.9% of respondents stated that there are secret organizations that have great influence on political decisions. Slightly more than 20% believe that politicians and other leaders are merely puppets of powers behind the scenes. "One in ten people in Germany believes that the Corona pandemic was caused by secret powers (9.8%). Even more people believe that compulsory vaccinations are being introduced through the pandemic (17.2%)" (Lamberty & Rees 2021, p. 293). Especially AfD voters (66.7%) and non-voters (57.6%) believe in conspiracies. The fewest conspiracy believers seem to be among supporters of Alliance 90/The Greens (7.2%) and the FDP (5.6%) (ibid., p. 295).

Jan-Willem van Prooijen and Nienke Böhm (2023) have, in two panel studies[2] (one in the Netherlands with 4500 people, the other in the USA with 600 people), drawn attention to an interesting connection between vaccine skepticism (or hesitant willingness to vaccinate) and conspiracy narratives (e.g., "Covid-19 is a hoax invented by interest groups for financial reasons"). Such conspiracy narratives are not always and everywhere the cause of vaccine skepticism. This skepticism can also be a reason to seek out conspiracy narratives with which one can justify one's own skepticism or fear of vaccinations to oneself and others.

Among the storytellers of conspiracy myths are the anti-vaxxers, the most extreme of whom see vaccination as a "bioweapon" deployed, as the case may be, by Rothschild, Bill Gates, the pharmaceutical industry, or—quite generally—by an elite seeking to control the world. The virus SARS-CoV-2, the cause of the Corona pandemic, is said to have been unleashed on the world by this shadow government in order to dominate it. Covid-19 is supposed to create a new world order; the virus was bred in a laboratory; it spreads via the new 5G networks; Bill Gates wants to forcibly vaccinate humanity in order to monitor it; the shadow government keeps children imprisoned in secret cellars, tortures them, and taps their blood to produce a serum for eternal youth. With such other conspiracy myths, we have, so to speak, reached the pinnacle of *absurdities* in the Corona crisis.

But it is not only the well-known "crazies" on the far-right side of the political spectrum who believe in conspiracy narratives and spread them on the street or in social media. Some left-wing activists, for example, also see secret services, bank and hedge fund managers, and EU politicians as forces at work, turning the

[2] In panel studies or longitudinal studies, in contrast to cross-sectional studies, individuals are surveyed or interviewed at several points in time or waves, and the results of the individual survey waves are compared. If, as is usually the case in panel studies, the same individuals are repeatedly asked the same questions over time, it is possible not only to analyze correlations but also changes and, under certain circumstances, causal processes.

wheels of the world clock in secret circles. It is therefore not surprising that an international research group led by Roland Imhoff, in a study with 104,253 people from 26 countries, found that conspiracy mentalities—after controlling for important sociodemographic variables such as gender, age, and education—are associated above all with far-right, but also with far-left convictions. While in countries such as Belgium, France, Germany, the Netherlands, Austria, Poland, and Sweden, conspiracy myths are more likely to be believed and spread by the political right, in Spain, Romania, or Hungary it is the far left (Imhoff et al., 2022). But what does this tell us?

In this context (but not only in this one), there are social science attempts to revive an "old" concept that—like authoritarianism—is also closely associated with the name Erich Fromm. Erich Fromm called it "group narcissism" or "collective narcissism." He developed this concept in 1964 ("The Heart of Man. Its Genius for Good and Evil," Fromm, 1999a; original: 1964) and expanded it in 1973 ("The Anatomy of Human Destructiveness," Fromm, 1999b; original: 1973). The foundations can already be found in 1955 ("The Sane Society," Fromm, 1999c; original: 1955). "Group narcissism has important functions. Above all, it promotes solidarity and internal cohesion within the group and facilitates its manipulation, since it appeals to narcissistic prejudices. Secondly, it is extremely important as an element that provides satisfaction to the members of the group, especially those among them who would otherwise have little reason to feel proud and valuable" (Fromm, 1999b, pp. 182 f.). Group narcissism can be an important social-psychological source of social, economic, and political discrimination and aggression toward other groups. The leader of a group plays an important role in this. He conjures and represents his own superiority and grandiosity, as well as that of his own group, in a prototypical way. Meanwhile, there are a number of international (quantitative) studies that refer more or less explicitly to group narcissism in the Frommian sense and suggest that group narcissism should not be confused with authoritarianism (e.g., Cichocka, 2016; Golec de Zavala et al., 2019). These appear to be two relatively independent phenomena. In this respect, group narcissism is also suitable as a counter-concept to the "libertarian authoritarianism" of Amlinger and Nachtwey (2022; see above).

Group narcissists are evidently concerned with the glorification, exaggeration, and idealization of the image of the group to which they feel they belong (the family, the community, the nation, "Germany first," the political party that doesn't quite get its way, etc.). In this way, the individual can also glorify and idealize their own self. Wherever group narcissism flourishes (e.g., in seemingly left-behind regions, in AfD strongholds, in demonstrations against the "Corona dictatorship," among the "turncoats 2.0" or the "election losers," and in some scientific communities), it provides the "frustrated," the "angry citizens," and their ideologues with the mindset (in plain terms: the patterns of feeling, thinking, and behavior) to compensate—regardless of social status—for unfulfilled feelings of greatness and uniqueness through conspiracy myths, group hostility, nationalism, anti-Americanism, or antisemitism.

In a completely different context, but not inappropriate here, Paul Feyerabend writes, among other things: "Considering the narcissistic chauvinism of science such an examination would seem to be more than reasonable" (Feyerabend, 1976, p. 390).

Setting aside some *absurdities* and asking what remains after the coronavirus pandemic. Well, first, the memories of school closures, contact bans, blocked park benches, closed restaurants, existential worries of freelance artists, and the dying in hospitals, nursing homes, and in one's own four walls, etc. And then there was the dispute over crisis management during the pandemic. On April 31, 2023, government representatives and opposition politicians argued fiercely, loudly, and emotionally in the German Bundestag over the final report on the health impacts of the pandemic on children and adolescents. Federal Health Minister Karl Lauterbach admitted mistakes and stated, among other things: "The school closures did not need to last this long" (German Bundestag, 2023).

Could the contact bans in retirement and nursing homes or the school closures have been avoided? Karim Bschir and Simon Lohse (2022) consider this possible. With a thoroughly serious perspective, they confront the (health) political measures and reactions during the coronavirus pandemic with Paul Feyerabend's epistemic pluralism. And they conclude, among other things, that a higher degree of pluralism in health policy and greater inclusion of non-scientific perspectives during the pandemic would have been desirable, both for epistemological and political reasons. Epistemic pluralism means an *equal* interplay of health policymakers, policy advisors, epidemiologists, social scientists, experts with local knowledge, nurses, pastoral caregivers, school experts, and parents, instead of one-sided elite discourses, when it comes to decisions for efficient and humane crisis management.

"The citizens are not ignorant. They live in a democracy in which information flows freely from one citizen to another. They do not just live; they participate in people's courts, public assemblies, cast their votes there, are called upon to judge plays written by the greatest dramatists in world history, are witnesses to public debates, athletic competitions, military conflicts. According to Protagoras (presumably 490 to 411 BC; WF), the knowledge that emerges from this unstructured but very free learning process is sufficient to judge all events in a democracy" (Feyerabend, in: Feyerabend & Thomas, 1985, pp. 327).

References

Adorno, T. W., Frenkel-Brunswick, E., Levinson, D. J., & Sanford, R. N. (1950). *The authoritarian personality*. Harper & Row.

AGENDA2030. (2023). https://www.2030agenda.de/de/ueber-uns. Accessed: June 26, 2023.

Amlinger, C., & Nachtwey, O. (2022). *Gekränkte Freiheit. Aspekte des Libertären Autoritarismus*. Suhrkamp.

BILDblog. (2020). Wie die „Bild"-Redaktion mit schmutzigen Tricks versucht, Christian Drosten zu zerlegen. https://bildblog.de/121365/wie-die-bild-redaktion-mit-schmutzigen-tricks-versucht-christian-drosten-zu-zerlegen/. Accessed: June 26, 2023.

BMWK. (2020). Informationen des Bundeswirtschaftsministeriums zur Aufhebung der Allgemeinverfügung für Schutzausrüstung. https://www.bmwk.de/Redaktion/DE/Pressemitteilungen/2020/20200319-informationen-zur-aufhebung-der-allgemeinverfuegung-fuer-schutzausruestung.html. Accessed: June 26, 2023.

Bschir, K., & Lohse, S. (2022). Pandemics, policy, and pluralism: A Feyerabend-inspired perspective on COVID-19. *Synthese, 200*(6), 441 ff.

Cichocka, A. (2016). Understanding defensive and secure in-group positivity: The role of collective narcissism. *European Review of Social Psychology, 27*(1), 283–317.

Decker, O., Kalkstein, F., Schuler, J., Celik, K., Brähler, E., Clemens, V., & Fegert, J. M. (2022). Polarisierung und autoritäre Dynamiken während der Pandemie. In O. Decker, J. Kiess, A. Heller, & E. Brähler (Eds.), *Autoritäre Dynamiken in unsicheren Zeiten* (pp. 91–126). Psychosozial-Verlag.

Deutscher Bundestag. (2023). Drucksache 20/5650. https://www.bundestag.de/dokumente/protokolle/amtlicheprotokolle/ap20098-945508. Accessed: Aug. 14, 2023.

Die Bundesregierung. (2020). 22. März 2020: Regeln zum Corona-Virus. https://www.bundesregierung.de/breg-de/leichte-sprache/22-maerz-2020-regeln-zum-corona-virus-1733310. Accessed: June 26, 2023.

Die Presse. (2020). Wie wirksam sind Masken? Forscher sind sich uneins. https://www.diepresse.com/5795101/wie-wirksam-sind-masken-forscher-sind-sich-uneins. Accessed: June 26, 2023.

Drosten, C. (2020). Prof. Dr. Drosten Ende Januar 2020 auf RBB—Mit einer Maske ist das Virus nicht aufzuhalten. https://www.bitchute.com/video/1x2upVssJtpC/. Accessed: June 26, 2023.

Drosten, C. (2022). Christian Drosten zur Corona-Lage in Deutschland: „Nach meiner Einschätzung ist die Pandemie vorbei". https://www.tagesspiegel.de/wissen/corona-experte-drosten-nach-meiner-einschatzung-ist-die-pandemie-vorbei-9089959.html. Accessed: June 29, 2023.

Feyerabend, P. K. (1976). Logic, literacy, and Professor Gellner. *The British Journal for the Philosophy of Science, 27*(4), 381–391.

Feyerabend, P. K. (1985). Die Rolle von Fachleuten in einer freien Gesellschaft. In P. Feyerabend & C. Thomas (Eds.), *Grenzprobleme der Wissenschaften.* (pp. 325–333). Eidgenössische Technische Hochschule, Verlag der Fachvereine.

Feyerabend, P. K. (2005a). *Die Vernichtung der Vielfalt. Ein Bericht,* herausgegeben von Peter Engelmann. Passagen Verlag.

Feyerabend, P. K. (2005b; Original: Farewell to Reason, 1987, pp. 49–62.). Notes on Relativism. In J. Medina & D. Wood (Eds.), *Truth—Engagements across philosophical traditions.* (pp. 146–156). Blackwell Publishing.

Forsa. (2021). Befragung von nicht geimpften Personen zu den Gründen für die fehlende Inanspruchnahme der Corona-Schutzimpfung. https://www.bundesgesundheitsministerium.de/ministerium/meldungen/ungeimpfte-wollen-sich-nicht-ueberzeugen-lassen.html. Accessed: June 29, 2023.

Frankfurter Rundschau. (2021). Nüßlein und Löbel: Maskenaffäre ist beispielhaft für das Politik-Verständnis von CDU und CSU. https://www.fr.de/meinung/kommentare/cdu-masken-affaere-nuesslein-loebel-union-kommentar-ideologie-korruption-90232540.html. Accessed: June 26, 2023.

Frei, N., Schäfer, R., & Nachtwey, O. (2021). Die Proteste gegen die Corona-Maßnahmen: Eine soziologische Annäherung. *Forschungsjournal Soziale Bewegungen, 34*(2), 249–258.

Frindte, W. (2022). *Quo Vadis, Humanismus?* Springer.

Fromm, E. (1999a; Original: 1964). Die Seele des Menschen. Ihre Fähigkeit zum Guten und Bösen. *Erich-Fromm-Gesamtausgabe in 12 Bänden, Band 2,* herausgegeben von R. Funk. Deutsche Verlags-Anstalt.

Fromm, E. (1999b; Original: 1973). Anatomie der menschlichen Destruktivität. *Erich-Fromm-Gesamtausgabe in 12 Bänden, Band 7,* herausgegeben von R. Funk. Deutsche Verlags-Anstalt.

Fromm, E. (1999c; Original: 1955). Wege aus einer kranken Gesellschaft. *Erich-Fromm-Gesamtausgabe in 12 Bänden, Band 4,* herausgegeben von R. Funk. Deutsche Verlags-Anstalt.

Golec de Zavala, A., Dyduch-Hazar, K., & Lantos, D. (2019). Collective narcissism: Political consequences of investing self-worth in the ingroup's image. *Political Psychology, 40,* 37–74.

Grande, E., Hutter, S., Hunger, S., & Kanol, E. (2021). Alles Covidioten? Politische Potenziale des Corona-Protests in Deutschland. WZB Discussion Paper, No. ZZ 2021–601, Wissenschaftszentrum Berlin für Sozialforschung (WZB). https://www.econstor.eu/bitstream/10419/234470/1/1759173207.pdf. Accessed: Jule 5, 2023.

Handelsblatt (2020). Vermummungspflicht könnte die nächste Verschärfung in Deutschland sein. https://www.handelsblatt.com/politik/deutschland/corona-pandemie-vermummungspflicht-koennte-die-naechste-verschaerfung-in-deutschland-sein/25699404.html. Accessed: June 26, 2023.

Hanefeld, S. M. (2022). *Der Corona-Skandal: Warum wir nicht alles glauben sollten.* Books on Demand.

Hasler, P. T. (2011). *Aktien richtig bewerten.* Springer Gabler.

Huffpost. (2020). Trump's Bonkers Coronavirus ‚Herd Mentality' Claim Lights Up Twitter. Quelle: https://www.huffpost.com/entry/donald-trump-herd-mentality_n_5f618e1ac5b6e27db134a42b. Accessed: July 12, 2023.

Hunger, S., Hutter, S., & Kanol, E. (2023). The mobilization potential of anti-containment protests in Germany. *West European Politics, 46*(4), 812–840.

Imhoff, R., Zimmer, F., Klein, O., António, J. H., Babinska, M., Bangerter, A., … & Van Prooijen, J. W. (2022). Conspiracy mentality and political orientation across 26 countries. *Nature Human Behaviour, 6*(3), 392–403.

Johns Hopkins University. (2020). https://coronavirus.jhu.edu/map.html. Accessed: Sept. 20, 2020.

Johns Hopkins University. (2023). https://coronavirus.jhu.edu/map.html. Accessed: June 26, 2023.

Kühne, S., et al. (2020): Gesellschaftlicher Zusammenhalt in Zeiten von Corona: Eine Chance in der Krise? *SOEPpapers on Multidisciplinary Panel Data Research, No. 1091,* Deutsches Institut für Wirtschaftsforschung (DIW), Berlin.

Lamberty, P., & Rees, J. H. (2021). Gefährliche Mythen: Verschwörungserzählungen als Bedrohung für die Gesellschaft. In A. Zick & B. Küpper (Eds.), *Die geforderte Mitte—Rechtsextreme und demokratiegefährdende Einstellungen in Deutschland 2020/21.* Friedrich-Ebert-Stiftung. Verlag Dietz Nachf.

Merkur.de. (2020). Corona in Italien: Tausende Ärzte und Krankenpfleger haben sich mit dem Coronavirus infiziert. https://www.merkur.de/welt/coronavirus-italien-tote-fallzahlen-infizierte-covid-19-news-lombardei-zr-13642289.html. Accessed: June 26, 2023.

Neumayer, E., Pfaff, K., & Plümper, T. (2021). Protest Against COVID-19 Containment Policies in European Countries. Available at SSRN: https://ssrn.com/abstract=3844989or.

Paul-Ehrlich-Institut. (2023). Sicherheit von COVID-19-Impfstoffen. https://www.pei.de/DE/newsroom/dossier/coronavirus/arzneimittelsicherheit.html. Accessed: June 26, 2023.

Pink, I., & Welte, T. (2023). Risiko und Häufigkeit von Long-COVID. *CME, 20*(1–2), 12–17.

Polanyi, K. (2019; Original: 1944). *The Great Transformation.* Suhrkamp.

Quent, M., Richter, C., & Salheiser, A. (2022). *Klimarassismus. Der Kampf der Rechten gegen die ökologische Wende.* Piper Verlag.

Redaktionsnetzwerk Deutschland. (2020). https://www.rnd.de/politik/corona-trump-gegen-seine-verschworung-der-fake-news-medien-S5PRBMJOS4ZH3YVTWHJBDOKP3U.html. Accessed: Oct. 28, 2020.

Reuband, K. H. (2022). AfD-Affinitäten, Corona-bezogene Einstellungen und Proteste gegen die Corona-Maßnahmen: Eine empirische Analyse auf Bundesländerebene. *Zeitschrift für Parteienwissenschaften (MIP), 28*(1), 67–94.

Richter, C., Wächter, M., Reinecke, J., Salheiser, A., Quent, M., & Wjst, M. (2021). Politische Raumkultur als Verstärker der Corona-Pandemie? Einflussfaktoren auf die regionale Inzidenzentwicklung in Deutschland in der ersten und zweiten Pandemiewelle 2020. *ZRex–Zeitschrift für Rechtsextremismusforschung, 1*(2), 191–211.

Richter, S., Faas, T., Joly, P., & Schieferdecker, D. (2022). Sag, wie hältst Du's mit der Impfpflicht?: Einstellungen der deutschen Bevölkerung zur Einführung einer allgemeinen Corona-Impfpflicht. *Rapid-Covid, Policy Brief*, 1/2022.

Robert Koch Institut. (2023). Wirksamkeit. https://www.rki.de/SharedDocs/FAQ/COVID-Impfen/FAQ_Liste_Wirksamkeit.html. Accessed: June 26, 2023.

Soldt, R. (2020). Die Organisationsstruktur hinter den „Hygiene-Demos". Frankfurter Allgemeine Zeitung, 4. August 2020. https://www.faz.net/aktuell/politik/inland/berlin-die-organisationsstruktur-hinter-den-corona-demos-16888674.html. Accessed: June 29, 2023.

Spöri, Z., & Eichhorn, J. (2021). Wer glaubt (nicht mehr) an Corona-Verschwörungsmythen. https://dpart.org/publications/wer-glaubt-nicht-mehr-an-corona-verschwo-rungsmythen/. Accessed: June 29, 2023.

Süddeutsche Zeitung. (2020). „Coronavirus" war auch bei Asterix der Böse. https://www.stuttgarter-zeitung.de/inhalt.nach-tod-von-albert-uderzo-coronavirus-war-auch-bei-asterix-der-boese.d32400ac-9386-4703-b8b9-9b3c26b74269.html. Accessed: June 26, 2023.

Tagesschau. (2020a). Trump sieht sich im Virus-Krieg, 19.03.2020. https://www.tagesschau.de/ausland/trump-coronavirus-109.html. Accessed: July 2, 2023.

Tagesschau. (2020b). Fernsehansprache zur aktuellen Lage in der Corona-Pandemie am 11.04. 2020. https://www.bundespraesident.de/SharedDocs/Reden/DE/Frank-Walter-Steinmeier/Reden/2020/04/200411-TV-Ansprache-Corona-Ostern.html. Accessed: July 2, 2023.

Trump, D. (2020). https://twitter.com/realDonaldTrump/status/1320708386476990464. Accessed: Oct. 28, 2020.

UN-Informationszentrum. (2020). UN stellt COVID-19-Plan vor, der „das Virus besiegen und eine bessere Welt schaffen" könnte. https://unric.org/de/covid-19-plan-guterres/. Accessed: June 26, 2023.

Van Prooijen, J. W., & Böhm, N. (2023). Do conspiracy theories shape or rationalize vaccination hesitancy over time? *Social Psychological and Personality Science*. https://doi.org/10.1177/19485506231181659. Accessed: July 12, 2023.

Watson, O. J., Barnsley, G., Toor, J., Hogan, A. B., Winskill, P., & Ghani, A. C. (2022). Global impact of the first year of COVID-19 vaccination: A mathematical modelling study. *The Lancet Infectious Diseases, 22*(9), 1293–1302.

Widmer, P. 2020. Wir sind nicht im Krieg. Brachiales Vokabular ist in der Corona-Pandemie en vogue. Das ist folgenreich. https://www.alexandria.unisg.ch/entities/publication/e3725da5-e664-47d4-b0db-78ab16ee21c2/details. Accessed: July 2, 2023.

Wielga, J. & Enste, P. (2022). Zwischen Angst, Skepsis und Verweigerung: Was wissen wir über Menschen mit Impfvorbehalten in der Covid-19-Pandemie? (No. 03/2022). Forschung Aktuell.

Zeit Online. (2020). AfD spricht von „Corona-Diktatur". https://www.zeit.de/politik/deutschland/2020-10/corona-beschluesse-angela-merkel-alexander-gauland-kritik-opposition. Accessed: July 2, 2023.

17 Climate Catastrophe

"But what can we do in a time like today, which has not yet reached equilibrium? […] The answer is simple: with a few exceptions, we will act barbarically: we will punish and kill, set war against war, teachers against students, 'spiritual leaders' against the public and against each other, we will speak in morally high-flown words about offenses and demand that violations of the law be prevented by force. But while we continue to live our lives in this way, we should at least try to give our children a chance. We should offer them love and security, not principles" (Feyerabend, 1995, pp. 237).

17.1 Shortly After Midnight?[1]

On July 16, 2023, it was raining at St. Finian's Bay. We were on vacation. Outside, the temperature was 14 degrees Celsius. In the pub, we mused with Paddy about the weather. Two hours later, the temperature rose to 18 degrees. The sun was shining. The weather was typically Irish. "May the wind be always at your back. May the sun shine warm upon your face and the rain fall soft upon your fields. And until we meet again, may God hold you in the palm of his hand," goes an old Irish blessing.

In the spring and summer of 2023, the forests of Canada were burning. Nine million hectares of land had gone up in smoke by early July. More than 150,000 people had to leave their homes (ZDF, 2023a). The smoke clouds drifted over

[1] Some of the arguments presented here I have already expressed elsewhere (Frindte, 2022, Chap. 20).

W. Frindte, *Against Narrow-Mindedness and Chauvinism—Journeying Through Absurd Times with Paul K. Feyerabend*,
https://doi.org/10.1007/978-3-658-50113-6_17

the east coast of North America and also reached Europe. There, it soon became hot as well. Temperatures in Southern Europe—in Italy, Greece, or Spain—were above 40, in some cases over 45 degrees Celsius in mid-July. In the Mediterranean region, firefighters battled terrible wildfires. People lost their homes and possessions. Some lost their lives. The climate change monitoring service Copernicus (Copernicus Climate Change Service) of the European Union reported that the first three weeks of July 2023 were the warmest three-week period worldwide since records began in 1940 (Copernicus, 2023). Then a polar front moved across Europe. It became cold, windy, and at times rained heavily. In Slovenia, Croatia, and Austria, there were severe storms and floods in August 2023. In Slovenia, the swollen rivers flooded two-thirds of the entire country. The *Tagesschau* asked at the beginning of August: "Is this still weather or already climate change?" (Tagesschau, 2023a). And *ZDF heute* answered: "July probably the hottest month in thousands of years" (ZDF, 2023b). "Climate change is here. It is frightening. And it is only the beginning," said UN Secretary-General António Guterres, warning of the clear and tragic consequences: "Children swept away by monsoon rains, families fleeing from flames (and) workers collapsing in scorching heat" (quoted in UNRIC—United Nations Regional Information Centre, 2023).

Since former US President Donald Trump is ran for reelection in 2024, let us recall this "stubborn child," as some of his former associates now call him: In mid-September 2020, 50 days before the presidential election, which he famously lost, Donald Trump visited California to see the catastrophe for himself. Devastating wildfires on the west coast of the USA (especially in California, Oregon, and Washington State, but also in Utah, Wyoming, and Arizona) had destroyed more than 12,500 square kilometers of land, forests, and about 4,500 buildings. Over 20,000 firefighters tried to contain the fires, mostly without success. Tens of thousands of people lost their homes, and over 500,000 people were asked to prepare for evacuation.

In a conversation at the time, Trump was pointedly reminded by California's Democratic Governor Gavin Newsom and the head of the California *Department of Conservation and Natural Resources* of the close connection between climate change and the catastrophic wildfires. Trump responded with the *narrow-minded* remark that it would soon get cooler, even if the scientists did not know that yet (NBC News, 2020). Furthermore, California should take care of better forest management, Trump added a day later in an interview with *Fox News*. For example, one could take Austria as a model. The people there, Trump said, live in forest cities with many more "explosive trees" than California, and yet people manage the fires. Trump in the original: "You look at countries, Austria, you look at so many countries. They live in the forest, they're considered forest cities. So many of them. And they don't have fires like this. And they have more explosive trees. They have trees that will catch easier, but they maintain their fire." It is easy to imagine that Austrians found it quite amusing to be described by Donald Trump as residents of forest cities (Huffpost, 2020). In June 2017, Trump had announced the withdrawal from the Paris Climate Agreement. He had described the man-made dangers of climate change as a *hoax* and fake news. After Joe Biden was elected as the new US President on November 3, 2020, the USA rejoined the circle of signatories to the Paris Agreement in February 2021. And now the stubborn child is back in power.

In April 2023, a majority of Americans advocated that the USA should become climate neutral by 2050. On average, 54% of all Americans surveyed said that climate change was a major threat. While 78% of people who lean toward the Democrats agreed, only 23% of Republican supporters did (Pew Research Center, 2023). In separate surveys, according to Pew Research, it was shown that concern about climate change is greater in other countries than in the USA. For example, 81% of French adults and 73% of Germans would perceive climate change as a major threat. Perhaps so.

In the monthly opinion poll conducted by *infratest dimap* (ARD-Deutschlandtrend, 2023), a total of 26% of respondents in April 2023 stated that environmental protection or climate change were the most important problems. Unlike Pew Research, however, *infratest dimap* presented the 1,300 respondents with an *open* question: "In your opinion, what is the most important political problem that German politics must address as a priority? And what is the second most important?" Be that as it may, in Germany, opinions on climate protection—depending on one's position in the political spectrum—are more than diverse. According to the ZDF Politbarometer, in June 2023 about 37% of respondents said that the previous measures for climate protection go too far. This was the view of 87% of AfD supporters, 44% of CDU/CSU supporters, 35% of FDP followers, 24% of SPD fans, 21% of the Left, and 3% of the Greens (Statista, 2023).

Regardless of political orientation, the majority of Germans—in spite of the Russian war of aggression against Ukraine and the still high inflation rates—considered climate change to be one of the most pressing social problems of our time in July 2023. This is the result of the *Social Sustainability Barometer,* an annual representative survey of more than 6,500 people across Germany on topics related to the energy and transport transition, conducted as part of the Ariadne project at the Helmholtz Centre Potsdam. When people are asked what bothers them most about the implementation of the energy transition, 53% highlight the *political management* of the processes (Sustainability Barometer, 2023). This will be discussed in more detail.

17.2 Of Deniers and Climate Racists

From July 29 to August 6, 2023, the AfD (Alternative for Germany) met in Magdeburg to gear up for the 2024 European elections and to deliberate on its election program. In the main motion presented for the program and then in the discussions at the party congress, the AfD not only called for a "confederation of European nations" to replace the existing European Union, a return to the Deutschmark, the closure of borders to protect against "illegal entries," the "immediate lifting of economic sanctions against Russia," and much more. The main motion also targets climate protection measures. "We do not share the irrational CO_2 hysteria that structurally destroys our society, culture, and way of life. [...] We do not want any EU limits that climate lobbyists can enforce against Germany's most fundamental interests" (AfD, main motion, 2023, p. 71). At the party congress in Magdeburg, master painter and varnisher Tino Chrupalla specified the accusation of "climate hysteria" and said: "1.5 degrees of global warming? Nothing will happen if the earth is 1.5 degrees warmer. These are all things where hysteria is being used. When there is talk of the melting of the poles: The North Pole, as we can see, is in good shape" (ZDF, 2023c). These are not new arguments; similarly *narrow-minded* views were already found in the 2019 European election program. Also in 2019, members of the AfD parliamentary group in the Bundestag questioned in a minor inquiry whether 97% of all climate scientists largely agreed on the fundamental findings regarding human-caused climate change. The German federal government, referring to several studies, clearly rejected the AfD's doubts in its response of August 23, 2019, and reaffirmed the view that "around 99% of scientists who publish specialist articles on climate protection are convinced that climate change is caused by humans" (German Bundestag, 2019, p. 5).

So why the excitement now? As is well known, in other European countries it is also the supporters of right-wing populist parties (e.g., Austria's FPÖ, Britain's UKIP, or the Dutch PVV) who dispute the scientific findings on climate change. Stella Schaller and Alexander Carius from the Berlin *think tank adelphi* have examined the voting behavior of right-wing and right-wing populist parties in the European Parliament on climate protection issues and conclude that two out of three right-wing populist members regularly vote against climate and energy policy measures. In this context, the AfD particularly acts as a representative of local initiatives that, for example, seek to prevent wind farms (Schaller & Carius, 2019).

Well, I have at least three reasons to get upset about the AfD: a general one, a particular one, and a global one. In *general*, in the spirit of Paul Feyerabend,

I consider AfD members to be cranks, to be crackpots (see Chap. 11). According to Feyerabend—as Jamie Shaw has shown (Shaw, 2021)—*cranks* are actors who only want to push through their own views and agendas and are not interested in questioning their opinions in light of other perspectives. These can be activists who immunize themselves against other viewpoints, or local experts who are unwilling to consider views that contradict their personal experiences. As a reminder: At a state party congress of the AfD in Brandenburg at the beginning of 2019, Alexander Gauland said he did not believe in the "fairy tale of man-made climate change" (Frankfurter Allgemeine, 2019). In the equally hot summer of 2018, Beatrix Amelie Ehrengard Eilika von Storch tweeted: "Yes. It is warm. Very much so. But this hysterical climate crisis screeching by the climate Nazis is truly unbearable. Even if we all walk instead of building cars, all become gender gagsters, and only eat broccoli: the sun doesn't care" (Von Storch, 2018). And at the end of July 2023, the born Duchess of Oldenburg wrote on Twitter: "Germany has been groaning for days under the #heatwave in the #heightofsummer: Today again 36 degrees in Berlin. 17 degrees in the morning, 19 degrees in the afternoon. That must be #climatechange" (Von Storch, 2023).

You can call it eccentric or *narrow-minded.* Perhaps such statements are simply propaganda to please and keep the followers in line. In any case, such statements are dangerous for our society. Incidentally, the authors of "The Authoritarian Personality" also observed the cranks, the crackpots (Adorno et al., -Brunswick 1950, 1995, p. 331). Carolin Amlinger and Oliver Nachtwey see in today's conspiracists, with their affinity for conspiracy thinking and prejudice, their modern-day counterparts (Amlinger & Nachtwey, 2022, p. 191).

In any case, I do not want to be governed by dangerous crackpots in *particular.* As is well known, this danger exists in Thuringia, Saxony, Saxony-Anhalt, and perhaps even at the federal level. At the beginning of July 2023, according to a representative survey by *Infratest dimap*, 34% of 1,193 respondents from Thuringia stated that if state elections were held next Sunday, they would vote for the AfD. The currently (in 2023) governing coalition in Thuringia of the Left, Greens, and SPD would receive only 35% approval in total (infratest dimap, 2023a). In the nationwide survey (early August 2023; infratest dimap, 2023b), the AfD achieved 21% approval, placing it second behind the CDU/CSU (with 27%) and ahead of the SPD (17%), Greens (15%), FDP (7%), and the Left (4%).

How did those in power respond during this period? At the end of July 2023, the Thuringian chairman of the CDU, Mario Voigt, said that the AfD, like the Greens, was a party of fear that tried to convince people we were on the brink of the apocalypse (Spiegel, 2023a). Thuringia's then-Minister-President

Bodo Ramelow considered it wrong to talk only about the AfD and lose sight of substantive policy (Die Zeit, 2023). CDU leader Friedrich Merz initially expressed more or less approval of dealing with the AfD at the municipal level, but after vehement criticism from his own ranks, he backtracked and wrote on Twitter (now X) that there would be no cooperation between the CDU and the AfD even at the municipal level (Tagesschau, 2023b). The then-Minister of Economic Affairs Robert Habeck blamed the weaknesses and lack of direction of the CDU for the AfD's surge in the polls (Spiegel, 2023b). And then-Chancellor Olaf Scholz did not believe in the lasting strength of the AfD (ZDF, 2023d).

To paraphrase a sentence by Paul Feyerabend: You simply cannot rely on politicians. They have their own interests, which color their interpretation of the evidence and the cogency of that evidence; they know very little, but pretend to know far more (see also: Feyerabend, 1980, p. 188).

Admittedly, the quoted reactions of those in power are somewhat simplistic. However, the impression that the political leaders elected by the people and paid from tax revenues were not fully aware of the seriousness of the situation was unavoidable in the summer of 2023. The governing parties did not seem to have a master plan against the rise of the AfD.

I am left with the reference to the *global* reason that so agitates me in view of the AfD's rise. Climate change, climate crisis, or climate catastrophe? It is not up to the observers to decide whether the climate crisis has the characteristics of a crisis or not. The only standard is the people who suffer from human-caused climate change. Vulnerable groups threatened by climate change are people over 65, the chronically ill, and children—at least in Europe. For millions of people in other regions of the world, human-caused climate change is a catastrophe. AfD members and probably their supporters do not see this, or they do not want to see it.

> "[...] how can the actions of humans who viewed themselves as the masters of nature and society and whose achievements now threaten to destroy both be reintegrated with the rest of the world?, [...] do human beings have the right to shape nature and cultures different from their own according to their latest intellectual fashions" (Feyerabend, 1987, p. 709).

A research team from the British University of Exeter, Nanjing University in China, and Wageningen University in the Netherlands has published studies indicating that the habitability of the Earth is primarily found in areas where the annual mean temperature is around 11 to 15 degrees Celsius. Most people live in regions where the annual average temperature ranges between six and 28 degrees Celsius. Currently, nine percent of the world's population (a little more than 600 million people) live outside the "human climate niche" of 11 to 15 degrees, although the definition of the climate niche is certainly controversial. What is clear, however, is that by the end of the century, if current policies do not change significantly, global warming could have increased by 2.7 degrees Celsius. Then, 22 to 39% of the world's population would live outside this "niche." This would affect areas that already frequently experience extreme temperatures, such as parts of the Arabian Peninsula, large parts of India, Indonesia, the Philippines, Thailand, Vietnam, the Sahel zone, etc. If it were possible to limit global warming to 1.5 degrees, then "only" five percent of people would be exposed to global warming outside the "climate niche" (Lenton et al., 2023; Xu et al., 2020). And for these people, climate change would still be catastrophic. Simply taking a vacation in Ireland is not an option for them. Meanwhile, German arms manufacturers are making hefty profits. In May 2023, for example, Rheinmetall announced that in the first quarter, operating profit had risen by 8 million EUR to 92 million EUR compared to the same period the previous year. Other arms manufacturers in Germany, Europe, and the USA are faring no worse. What could be accomplished with this money to address human-caused climate change?

The studies on the "human climate niche" have not gone unchallenged. For example, the conclusions have been described as "alarmist" (Milkau, 2022, p. 81), or the assumption has been criticized that already nine to ten percent of the world's population exist outside the "climate niche." It is more likely, they argue, that currently everyone is living under livable conditions (Tol, 2023, p. 7). If such statements were not so *absurd*, one could sit back in the professor's chair and think, to err is human. In the case of human-caused climate change, however, error could be fatal.

Whether the population accepts human-caused climate change or not, whether people support or reject climate-friendly measures, whether they do so for political or financial reasons, what is striking are the mostly escalating debates between opponents and proponents. The disputes are carried out in popular science books, in magazines, on television, in sessions of the Bundestag and state parliaments, and primarily on online platforms. Among the opponents and deniers of human-caused climate change are pseudo-scientists,[2] journalists, politicians,

[2] Of course, women among the climate change deniers are also included.

YouTube celebrities, conservative think tanks, and—as already mentioned—right-wing populist and far-right organizations and parties. The climate change deniers send emotionally charged, angry tweets and posts (Hemsley et al., 2021), question scientific facts, spread false reports, and believe in conspiracy myths. Common arguments of climate change deniers include statements such as: the Earth is not warming at all; it is warming, but not due to human activity; something could be done about global warming, but it is too expensive or it is already too late; first, the countries that emit more CO_2 than Germany should be held accountable, etc. (Luczak, 2020, pp. 17 f.). Conspiracy myths fit well into the portfolio of climate change deniers, as the threatening consequences of human-caused climate change can conveniently be reinterpreted as another, perhaps lesser, threat: the threat posed by secret, conspiratorial decisions made by individuals or small groups with illegitimate intentions (see also Biddlestone et al., 2022). Thus, in the conspiracy myths of climate change deniers, it is not uncommon to hear whispers of the public being misled by secret agreements between climate researchers and politicians. The ultimate goal of such agreements, they claim, is to overthrow society and the state and establish a green "climate dictatorship."

Who are they, the climate change deniers and opponents, and what drives them?

Deep in the northwest of the USA, in Idaho, Kristin Haltinner and Dilshani Sarathchandra (2021) conducted a small interview study in which they asked 33 adult white Americans about the reasons for their skepticism toward human-caused climate change. The study is, of course, not representative, but it does document a narrative that could probably also be found in Germany. Some respondents cited religious reasons; for example: God will take care of it; or: climate change, floods, etc., are signs of the coming end of the world. Other interviewees rejected mainstream media and considered the hype about climate change to be a conspiracy by the United Nations and its allies.

Quantitative analyses suggest that among climate change skeptics and deniers, there is a disproportionately high number of religiously and conservatively minded older men with higher incomes compared to non-skeptics. Thus, it is not far-fetched to speak of a "conservative white male effect" (McCright & Dunlap, 2013). However, it should be noted that most of these studies were conducted in the USA. Therefore, it is reasonable to assume that there is a particular political culture in the USA that encourages citizens to evaluate statements about the climate and climate change more strongly through an ideological and worldview lens. That the United States is something special in this regard is hardly unknown and is also reflected in intercultural comparative analyses. Matthew J. Hornsey, Emily A. Harris, and Kelly S. Fielding (2018) compared the USA with 24 other countries (e.g., Argentina, Brazil, China, Germany, Great Britain, India, or Poland) and found that the correlations between skepticism toward human-caused climate change and ideological beliefs are stronger and more consistent in the USA than in any other country.

We therefore need to dig deeper to learn more about the skeptics and deniers in Europe and Germany. For this reason, Wouter Poortingaa and colleagues (2019) analyzed data from the *European Social Survey* (ESS) of over 44,000 people from 23 countries (including Russia).[3] It was found that the relationships between sociodemographic variables (e.g., age, gender, education), political orientations, and attitudes toward human-caused climate change vary greatly between different European regions. While men from Eastern and Northern Europe (e.g., Poland, Russia, Slovenia, or Norway and Sweden) cared less about climate change than women, such a gender effect was not observed in Western European countries (e.g., Belgium, Germany, and France). Here, in Western European countries, an age effect is observed; the older the respondents in these regions, the more skeptical they are about climate change.[4]

In all countries, there are also clear, significant correlations between attitudes toward climate change and political orientations. People who place themselves more to the right on the political spectrum are less interested in the well-being of others and express more skepticism toward anthropogenic climate change. Meta-analyses also confirm the links between climate skepticism and political attitudes. Samantha Stanley and Marc Wilson (2019) analyzed 53 studies and show, among other things, that people with pronounced authoritarian beliefs have a more negative attitude toward the environment than those who are less authoritarian. Meta-analyses also show that climate-friendly behavior is strongly linked to trust in science and environmental organizations (Cologna & Siegrist, 2020). In order to foster this trust, however, environmental activists and environmental researchers must be perceived as impartial authorities, which is, as is well known, no simple matter.

---

[3] The ESS is one of the most renowned comparative studies, with which, for twenty years, citizens from Europe and Israel have been regularly surveyed about their opinions on social and political issues.

[4] However, Steffen Mau, Thomas Lux, and Linus Westheuser are unable to demonstrate a generational divide in Germans' climate awareness. "In fact, the older cohorts even express greater fears than the younger generations. In the group of those over 70, a full 85% are very worried about climate change, among the 50- to 59-year-olds this figure drops to 76%, and among the 16- to 29-year-olds it is only 62%" (Mau et al., 2023, pp. 217). So I am in good company.

Whether anthropogenic climate change is considered significant at all, is perceived by people, and motivates climate- and environmentally friendly behavior depends not least on which groups and social authorities people identify with, or what role climate change plays in the pursuit of a positive social identity. A person's social identity is that part of the self-image that derives from affiliations with various social groups and is linked to evaluations and emotions regarding these groups (Tajfel & Turner, 1979). Social identities play an important role, not least in the political debate about climate change. Members of political parties or political movements define themselves through their affiliation with these groups and base their sometimes opposing views on these affiliations. Some see themselves as climate change skeptics, who perceive their values and beliefs as threatened by the others, the climate activists, while the latter in turn perceive the skeptics as a threat to humanity (Mackay et al., 2021).

Humans have become a force of nature, one that may ultimately lead to their own demise. Climate change, global warming, species extinction, acidifying oceans, the concentration of greenhouse gases in the atmosphere, water scarcity, all kinds of toxins rapidly destroying the environment through industrialization, agriculture, and private households, etc.—all these are indicators of a transformation caused by humans themselves, by which they are now threatened. The problems brought about by climate change are among the greatest threats and challenges. Global climate change and global warming are largely man-made and threaten humanity in all parts of the world. Radical observers of human-induced transformation of the world, such as *Jason W. Moore,* an American historian and sociologist, argue, however, that it is not humanity or we humans who are responsible for climate change and environmental destruction, but modern capitalism. *Jason Moore* (2020) therefore speaks of the "Capitalocene." Matthias Quent, Christoph Richter, and Axel Salheiser (2022) argue similarly when they write that the history of anthropogenic climate change is the history of the "success model of capitalism."

If the history of anthropogenic climate change is a history of capitalism, then on the one hand, it is a success story. Consider, for example, the advances associated with modernity in terms of health, child mortality, life expectancy, liberalization of values, etc. But such a view of capitalism is "… a white, Western, mostly male, and both materially and culturally privileged" one (Quent et al., 2022, p. 23). If one changes perspective, then capitalism and its consequences, including anthropogenic climate change, are a history of exploitation, oppression, and severe inequalities. So it is not simply a matter of a climate capitalism that needs to be criticized. The denial of anthropogenic climate change amounts to racist supremacy. "Climate racism, in this sense, describes at a structural level the externalization of the ecological costs of industrial prosperity in the predominantly white West at the expense of predominantly non-white regions and people. Furthermore, climate racism describes the ideological and strategic backgrounds of the right's responses to the consequences of climate change and to demands for climate justice" (Quent et al., 2022, p. 27). It is not the "Occident" that is facing collapse, as the radical right propagates, "… but many Pacific islands are in fact being sunk into the sea by the Western model of industrial society" (ibid., p. 51). The model of Western industrial society, of the Christian-influenced "Occident," of enlightened democracies, of white supremacy, of the world order of inequality, of male domination is destroying humanity in the global South. That is strong stuff, but it does quite aptly capture the *chauvinism* of climate deniers and climate racists.

This is echoed by a remark from Paul Feyerabend in a footnote in "Creativity: A Dangerous Myth." Feyerabend refers to a work by Grazia Borrini comparing "rich" and "poor" countries (Borrini, 1985) and writes: "More recent studies of what is called, somewhat euphemistically, »developmental aid« have revealed the material and spiritual damage caused by Western interventions. The expansion of Western civilization robbed indigenous peoples of their dignity and their means of survival. War, slavery, simple murder seemed to be the right ways of dealing with »primitives«. But the humanitarians did not fare better than the gangsters. Imposing their own views of what it means to be human and how scarcity can be avoided, they often added to the destruction wrought by their predecessors" (Feyerabend, 1987, p. 709).

Certainly, one can debate whether "global North" and "global South" are helpful categories for analyzing and describing the global contradictions of the climate crisis, postmodern capitalism, and right-wing attacks on democracy. In serious scientific debates, there are indeed controversial positions on this. It is trivially agreed that these categories do not describe geographic entities. There is more disagreement about the extent to which these terms have the potential to strengthen the positions of various social actors from the global South in global networks of power (e.g., Kloß, 2017), or whether the designation "global South" conveys new forms of discrimination (e.g., da Silva, 2021). If, on the other hand, one looks at statistics on global wealth distribution, countries such as Switzerland, the USA, the so-called Special Administrative Region of Hong Kong,[5] Australia, New Zealand, Denmark, Canada, the Netherlands, Sweden, Belgium, and Singapore are the eleven richest countries. The countries in Africa are found far behind at the bottom of the list (Global Wealth Report, 2022). In this respect, the rough framework used by Matthias Quent, Christoph Richter, and Axel Salheiser to denounce global climate injustices or the climate policy of the North to the detriment of the South does work.

"Be fruitful and multiply and fill the earth and subdue it," it says in Genesis (1 Moses, 1:28). If one takes *Moses* at his word, it does not mean to subdue the earth in order to renew it at some point. The order is reversed: Renew, fill the earth, and then you may subdue it. This is not only what Moses says, but this is the commandment for humanity: "There is neither a commandment nor a license to destroy the earth […]. The earth is to be protected from abuse because it was created for the joy of mankind. To abuse the earth is not an ecological crime. It is a crime against humanity" (Chighel, 2020, p. 278).

17.3 что делать? Psychological Reflections

Grazia Borrini-Feyerabend and Rosemary Hill ask, among other things, how we humans must respond to human-induced climate change, the loss of biological diversity, and environmental disasters. One possible answer, according to the authors, could be adaptive governance, a learning-based management. This means, among other things, giving the relevant institutions, but also the people involved, the time and opportunity to learn and to incorporate both individual and collectively shared knowledge as well as individual and collective values and norms into the respective decisions (Borrini-Feyerabend & Hill, 2015, p. 196).

Looking at the present in the year 2023, however, one must ask whether management in politics and science is given the time to learn and to what extent the knowledge, values, and norms of those involved are taken into account in dealing with human-induced climate change. Certainly, these are primarily political questions that must be decided at the political level and within civil society.

For some, things are moving too slowly, for example for the *Last Generation,* who claim we have only two to three years left to leave the "fossil path of destruction" (Last Generation, 2023). Others are calling for a ban on the

[5] Which, as is well known, has long since ceased to be able to govern itself.

organization *Last Generation,* because they see it as a threat to the constitutional order of the Federal Republic (AfD motion to the German Bundestag, 2023). Still others, such as several politicians from the CDU/CSU, want the Last Generation to be "only" monitored by the Office for the Protection of the Constitution (Tagesschau, 2023c). The government of the Federal Republic criticized the "massive disruptions of public order" by activists of the *Last Generation.* Chancellor Scholz called the Last Generation's gluing actions "completely crazy" (Frankfurter Allgemeine, 2023). It should be noted that activists of the Last Generation not only call for protest, but regularly try to draw attention to the climate catastrophe through sit-ins, smearing museum artworks, or gluing themselves to roads.

According to a survey by Südwestrundfunk, in June 2023, 85% of respondents considered the actions (especially blocking roads) of climate activists to be "not justified" (Tagesschau, 2023d). Protest researcher Dieter Rucht commented on the civil disobedience of the *Last Generation* and said: "In view of the inadequate political response to the challenges of climate policy, I consider civil disobedience—alongside all other, but so far hardly effective, means of influence—to be necessary. However, it should not be oriented only towards the goals, but also in terms of the concrete protest practice, not towards the allure of spectacle and the heroization of 'resistance,' but should, with the act of disruption, primarily confront those politically responsible (instead of uninvolved third parties), take on the effort of continuous persuasion, and be embedded in a broader critique of social conditions" (Rucht, 2023, p. 25). I also do not find the actions of the *Last Generation* particularly effective. However, they have my sympathy for one reason: They are not authoritarian nationalists like the Reichsbürger. They see themselves as part of humanity globally threatened by human-induced climate change.

On some current findings: I will limit myself to my social-psychological expertise: Social psychology has indeed produced an impressive amount of research on how people deal with climate change. People who perceive human-induced climate change not only as an abstract message, but are sensitive enough to anticipate consequences affecting themselves *and* see themselves as part of a *global humanity* (global identity), are also more willing to engage in environmental protection, climate-friendly energy, and a livable future (e.g., Loy & Spence, 2020; Masson & Fritsche, 2021). Identification with nature (in the sense of a close connectedness to nature) is also—how could it be otherwise—closely linked to environmentally responsible behavior. Similar to identification with social groups and environmental movements, identification with nature can promote climate- and environmentally friendly behavior (Mackay & Schmitt, 2019). Ultimately, it is not just about heavy rainfall in southern Germany or dry soils in Mecklenburg-Western Pomerania, nor just about melting glaciers in the Alps, floods in Slovenia, Croatia, and Austria, or heat waves across Europe; it is not about this or that and there, but about real, worldwide threats to humanity as a whole.

Of course, human-induced climate change and its catastrophic consequences can cause fear, but also motivate individual environmentally friendly behavior. For example, Kim Pong Tam and colleagues found in a cross-country study in China, India, Japan, and the USA, among other things, that fear of climate change is more pronounced in the Chinese and Indian populations than in Japan and the USA, but across all four countries, it is positively associated with engagement in climate protection (especially regarding sustainable nutrition and climate activism; Tam et al., 2023). Motivational impulses from fear in the face of climate change for environmentally friendly actions can also be demonstrated, especially for younger age groups in the UK (Whitmarsh et al., 2022). When people identify with climate-friendly and action-oriented groups, not only can individual fear be reduced, but motivation for collective environmental and climate action can also be promoted (Stollberg & Jonas, 2021). Human-induced climate change requires collective action, and climate-friendly behavior is only effective in community with others (Fritsche et al., 2018). That is plausible and hardly trivial.

But what about the acceptance of political decisions regarding climate change and the motivation to behave and act in a climate-friendly way as individuals or as members of social groups? A meta-analysis by Magnus Bergquist and colleagues provides initial insights (Bergquist et al., 2022). The psychologists and political scientists from Sweden asked to what extent people from different professional groups (e.g., farmers, scientists, politicians) from different regions (Asia, Europe, North America, Oceania) accept market-based and political instruments (e.g., introduction of CO_2 taxes). For this, 89 datasets with a total of 119,465 people from 33 countries were analyzed. Political beliefs (such as perceived fairness and assessment of the efficiency of economic and political instruments in dealing with climate change), assessment of climate change (e.g., worries, fears, and risk perception regarding climate change), psychological factors (e.g., altruism, selflessness or selfishness and egocentrism, the need for self-actualization, political left-right orientation), and sociodemographic characteristics (education, age, income, sexual orientation) were examined. Across the various geographic regions, these conditions naturally played different roles. Overall, and across all datasets, it was found that sexual orientation (gender) does not appear to play a substantial role in the assessment of climate-friendly economic and/or political decisions. Age, income, and education also have only a minor effect. The psychological factors vary in their influence on the acceptance of climate-friendly economic and political instruments. Particularly effective seem to be people's worries, fears, and risk assessments when it comes to accepting climate-friendly instruments. Most strongly, however, acceptance is influenced by the perceived fairness and the assessment of the efficiency of economic and political instruments. *That is, the fairer economic and political decisions to curb climate change are, and the more effective and useful they are perceived to be, the more likely far-reaching climate change instruments will be accepted.*

The usefulness of climate-friendly decisions is also highlighted by numerous other studies, for example, the one by Emily Schulte and colleagues (2022). This is also a meta-analysis, in which initially 110 international studies were evaluated, and finally, based on eight studies (with 1,714 participants), a model was developed to predict the motivation or intention to install photovoltaic systems in private households. The basis of this model is the theory of planned behavior by Icek Ajzen and Martin Fishbein (for an overview: Ajzen 1991; Fishbein & Ajzen, 2011).

Fishbein and Ajzen assume that it is easier to derive a specific intention to perform a behavior from a particular attitude than to predict the actual behavior itself. This is quite understandable. Suppose you have a positive attitude toward photovoltaic systems and intend to acquire such a system. However, you are prevented from actually carrying out this behavior because there are currently no technicians available to help you. This or another obstacle is difficult to foresee in an attitude theory. Therefore, it makes sense to first distinguish between behavior and behavioral intention. *Behavioral intention* refers to the intention to perform a specific behavior at some more or less precisely determined point in the future. A behavioral intention depends on a person's attitude toward the intended behavior, the subjective norm, and perceived behavioral control. The *attitude* toward a behavior refers to the evaluation by which a person judges a particular behavior positively or negatively. For example, it makes a difference whether you want to buy a photovoltaic system because you believe such a system is environmentally friendly, or whether you intend to buy one simply because these systems are currently trendy. A second component is the so-called *subjective norm.* This refers to the social pressure perceived by a person to perform or not perform a particular behavior. As the name suggests, the subjective norm is not an actual social norm to do or refrain from something, but rather the subjective perception and interpretation of such a social norm. To stay with our example, the question could be: How convinced are you, for instance, that your friends and acquaintances expect you to acquire a photovoltaic system; what do you know about the supposed expectations of others regarding your climate-friendly actions? The third component in the theory of planned behavior is *perceived behavioral control.* Perceived behavioral control refers to experience-based knowledge about the conditions that can hinder or facilitate one's own behavior. If a person has sufficient information about such conditions, it can be assumed that they are able to adequately assess their own control over their behavior. Depending on, to return to the above example, whether you believe you can decide for yourself to acquire such a system (or alternatively a heat pump), or whether you are being forced to do so, you will perceive your control over your behavior (buying a photovoltaic system or not) differently.

Returning to the study by Emily Schulte and colleagues. Without going into the results in detail, it can certainly be said: Sociodemographic characteristics, such as age or sexual orientation, again seem to be less relevant when asking whether people are willing or encouraged to acquire climate-friendly energy generation systems. Much more decisive are likely to be a) attitudes toward the advantages and disadvantages of such systems, b) the perception of one's own behavioral control, that is, the assessment of whether one is the master of one's own decisions or whether one is being pushed into such climate-friendly purchases (by politics, business, or the media), and finally c) how strong the social pressure is from friends, acquaintances, or other people with whom one identifies, to finally heat one's own house in a climate-friendly way. And, of course, the whole thing is also a question of money.

If climate-friendly action is considered a task to be solved, then certainly other conditions come into play that determine how strong the intention is to make positive contributions to environmental and climate protection and to accept corresponding political requirements (see Fig. 17.1).

Climate-friendly action encompasses individual and/or collective actions that can be taken to address human-caused climate change, for example: supporting climate protection projects, heating efficiently, climate-friendly nutrition, etc. Is it only about photovoltaic systems in private households and heat pumps, or also about general speed limits on highways, wind and solar plants in the region, and so on? The *importance of a climate-friendly action (CFA),* whether it is performed individually or collectively, *communally,* the *degrees of freedom* in its execution, the *feedback* after execution, the *perceived fairness* and the *efficiency* of the CFA are significant (e.g., Bergquist et al., 2022; Fritsche et al., 2021; Katz et al., 2022; Lou & Li, 2023). That is, among other things, to what extent the climate-friendly action affects one's own life and that of others, to what extent the actors have sufficient freedom of choice to perform a climate-friendly action themselves or whether they are forced to do so, how fair and just a CFA is perceived to be for the life design of individuals or groups, and how efficient one's

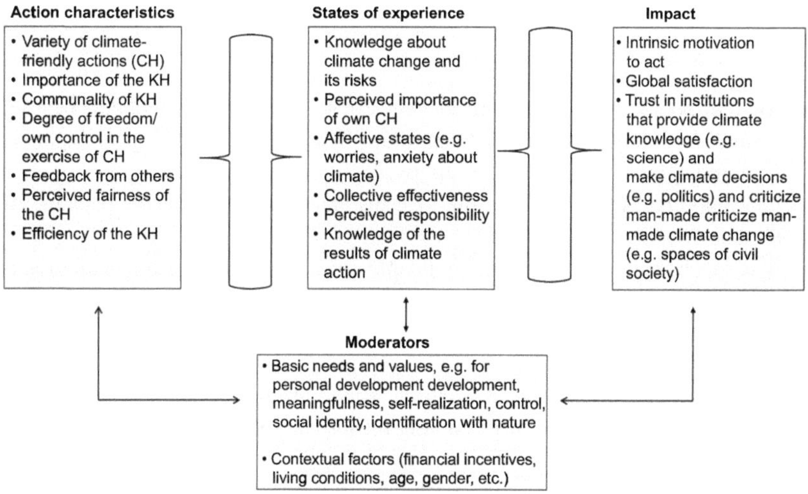

Fig. 17.1 Illustration of parameters for climate-friendly action

own action is assessed to be. These and other action characteristics influence various experiential states of the actors and are influenced by them. These include *knowledge* about human-caused climate change, the experienced *significance* of one's own actions, *affective states* such as climate anxiety and risk perception, but also the self-experienced and perceived *responsibility* for human-caused climate change, as well as the experience and *knowledge* that one's own actions are associated with (preferably positive) results. If climate-friendly action is to be self-organized and *intrinsic,* that is, motivated by one's own initiative, then this depends not least on precisely these action characteristics and experiential states. Through these pathways, *trust* in institutions can also be fostered, whose task it is to provide knowledge about climate change, make political decisions, or organize actions for climate change. Finally, climate-friendly action also depends on basic needs and values; these needs and their satisfaction, in turn, influence climate-friendly action. To what extent can people, through self-organized and/or politically mandated climate-friendly action, realize themselves, experience themselves as significant, achieve self-actualization, exercise control, identify with like-minded people and with nature? Contextual factors such as financial incentives, living conditions, age, and gender also influence climate-friendly action to varying degrees.

From a psychological perspective, one could therefore ask, among other things,

- whether people who, starting in January 2024, have to purchase a new heating system, have any options other than those based on renewable energies, aside from transitional periods and exceptions,
- how important these new heating systems are within the possible package of climate-friendly actions,
- how much freedom and autonomy people are granted by the government regulations, the new laws, and their consequences,
- whether restrictive laws might not trigger new affective states, such as financial worries, resistance, etc.,
- to what extent government regulations aimed at curbing human-induced climate change are capable of promoting intrinsically motivated climate-friendly behavior,
- whether, with these legal stipulations, basic needs such as personal development, self-actualization, etc., can be satisfied or are rather restricted?

I will refrain from providing answers, but I claim: The current governmental and political regulations are not always particularly motivating for climate-friendly action by individuals or groups. As is well known, communication about climate change between the various actors leaves much to be desired (see also Wolling et al., 2023), not to mention the cross-party disputes. Perhaps it would sometimes help to pay a little more attention to what people are actually saying—not to pander to them, but to communicate better with them. Anything else would be narrow-minded. As Paul Feyerabend put it: *"Not rationalist standards, not religious convictions, not humane impulses, but citizens' initiatives are the filter that separates useful from useless ideas and measures"* (Feyerabend, 1980, p. 77; emphasis in the original). Politicians should keep this in mind from time to time. But I admit, it is the difficult thing that is hard to do, even in a democratically constituted society.

References

Adorno, T. W., Frenkel-Brunswick, E., Levinson, D. J., & Sanford, R. N. (1950). *The Authoritarian Personality.* Harper & Row.

Adorno, T. W., Frenkel-Brunswick, E., Levinson, D. J., & Sanford, R. N. (1995). *Studien zum autoritären Charakter.* Suhrkamp.

AfD, Leitantrag der Bundesprogrammkommission. (2023). https://www.afd.de/magdeburg2023/. Accessed: Aug. 7, 2023.

AfD-Antrag an den Deutschen Bundestag. (2023). https://dserver.bundestag.de/btd/20/067/2006702.pdf. Accessed: Aug. 12, 2023.

Ajzen, I. (1991). The theory of planned behavior. *Organizational Behavior and Human Decision Processes, 50*(2), 179–211.

Amlinger, C., & Nachtwey, O. (2022). *Gekränkte Freiheit. Aspekte des Libertären Autoritarismus.* Suhrkamp.

ARD-Deutschlandtrend. (2023). Klimawandel als wichtigstes Problem. https://www.tagesschau.de/inland/deutschlandtrend/deutschlandtrend-3339.html; Accessed: Aug. 6, 2023.

Bergquist, M., Nilsson, A., Harring, N., & Jagers, S. C. (2022). Meta-analyses of fifteen determinants of public opinion about climate change taxes and laws. *Nature Climate Change, 12*(3), 235–240.

Biddlestone, M., Azevedo, F., & van der Linden, S. (2022). Climate of conspiracy: A meta-analysis of the consequences of belief in conspiracy theories about climate change. *Current Opinion in Psychology, 46,* 101390.

Borrini, G. (1985). Health and development: A marriage of heaven and hell? *Studies in third world society.* Univ. of Texas.

Borrini-Feyerabend, G., & Hill, R. (2015). Governance for the conservation of nature. *Protected Area Governance and Management, 7,* 169–206.

Chighel, M. (2020). *Kabale. Das Geheimnis des Hebräischen Humanismus im Lichte von Heideggers Denken.* Vittorio Klostermann GmbH.

Cologna, V., & Siegrist, M. (2020). The role of trust for climate change mitigation and adaptation behaviour: A meta-analysis. *Journal of Environmental Psychology, 69.* https://doi.org/10.1016/j.jenvp.2020.101428.

Copernicus. (2023). July 2023 sees multiple global temperature records broken. https://climate.copernicus.eu/july-2023-sees-multiple-global-temperature-records-broken#:~:text=The%20hottest%20day%20was%206,three%2Dweek%20period%20on%20record. Accessed: Aug. 6, 2023.

Da Silva, J. T. (2021). Rethinking the use of the term ‚Global South‘ in academic publishing. *European Science Editing, 47,* e67829.

Deutscher Bundestag. (2019). Drucksache 19/12631. https://dserver.bundestag.de/btd/19/126/1912631.pdf. Accessed: Aug. 7, 2023.

Die Zeit. (2023). Ramelow: Dauerdebatte über AfD nutzt den Rechtspopulisten. https://www.zeit.de/news/2023-08/06/ramelow-dauerdebatte-ueber-afd-nutzt-den-rechtspopulisten; Accessed: Aug. 7, 2023.

Feyerabend, P. K. (1980). *Erkenntnis für freie Menschen.* Suhrkamp.

Feyerabend, P. K. (1995). *Zeitverschwendung.* Suhrkamp.

Feyerabend, P. K. (1987). Creativity: A dangerous myth. *Critical Inquiry, 13*(4), 700–711.

Fishbein, M., & Ajzen, I. (2011). *Predicting and changing behavior: The reasoned action approach.* Taylor & Francis Group.

Frankfurter Allgemeine. (2019). AfD macht SPD vor Landtagswahl zum Hauptgegner. https://www.faz.net/aktuell/politik/inland/afd-brandenburg-macht-spd-vor-landtagswahl-zum-hauptgegner-15973969.html. Accessed: Aug. 7, 2023.

Frankfurter Allgemeine. (2023). „Letzte Generation" ist „fassungslos" über Scholz' Kritik. https://www.faz.net/aktuell/politik/inland/scholz-kritisiert-letzte-generation-aktivisten-fassungslos-18912714.html. Accessed: Aug. 12, 2023.

Frindte, W. (2022). *Quo Vadis, Humanismus?* Springer.

Fritsche, I., Barth, M., & Reese, G. (2021). Klimaschutz als kollektives Handeln. In L. Dohm, F. Peter, & K. v. Bronswijk (Eds.), *Climate Action – Psychologie der Klimakrise* (pp. 229–250). Psychosozial-Verlag.

Fritsche, I., Barth, M., Jugert, P., Masson, T., & Reese, G. (2018). A Social Identity Model of Pro-Environmental Action (SIMPEA). *Psychological Review, 125*(2), 245–269.

Global Wealth Report. (2022). https://www.credit-suisse.com/about-us/en/reports-research/global-wealth-report.html. Accessed: Aug. 8, 2023.

Haltinner, K., & Sarathchandra, D. (2021). The nature and nuance of climate change skepticism in the United States. *Rural Sociology.* https://doi.org/10.1111/ruso.12371.

Hemsley, J., Hopek, J., & Lee, J. (2021). Climate change deniers spread negative emotion laden tweets. https://ssrn.com/abstract=3837367. Accessed: Aug. 8, 2023.

Hornsey, M. J., Harris, E. A., & Fielding, K. S. (2018). Relationships among conspiratorial beliefs, conservatism and climate scepticism across nations. *Nature Climate Change, 8*(7), 614–620.

Huffpost. (2020). Austrian minister to Trump: No, We Don't Live In ‚Forest Cities,' Have exploding trees. https://www.huffpost.com/entry/austrian-minister-responds-trump-forest-cities-exploding-trees_n_5f63b7cac5b61845586 85df8. Accessed: July 12, 2023.

Infratest dimap. (2023a). ThüringenTREND Juli 2023. https://www.infratest-dimap.de/umfragen-analysen/bundeslaender/thueringen/laendertrend/2023/juli/. Accessed: Aug. 7, 2023.

Infratest dimap. (2023b). ARD-DeutschlandTREND August 2023. https://www.infratest-dimap.de/umfragen-analysen/bundesweit/ard-deutschlandtrend/2023/august/. Accessed: Aug. 7, 2023.

Katz, I. M., Rauvola, R. S., Rudolph, C. W., & Zacher, H. (2022). Employee green behavior: A meta-analysis. *Corporate Social Responsibility and Environmental Management, 29*(5), 1146–1157.

Kloß, S. T. (2017). The Global South as subversive practice: Challenges and potentials of a heuristic concept. *Global South, 11*(2), 1–17.

Lenton, T. M., Xu, C., Abrams, J. F., Ghadiali, A., Loriani, S., Sakschewski, B., & Scheffer, M. (2023). Quantifying the human cost of global warming. *Nature Sustainability, 1–11,*. https://doi.org/10.1038/s41893-023-01132-6.

Letzte Generation. (2023). Jetzt erst recht! https://letztegeneration.org/. Accessed: 12. Aug. 2023.

Lou, X., & Li, L. M. W. (2023). The relationship of environmental concern with public and private pro-environmental behaviours: A pre-registered meta-analysis. *European Journal of Social Psychology, 53*(1), 1–14.

Loy, L. S., & Spence, A. (2020). Reducing, and bridging, the psychological distance of climate change. *Journal of Environmental Psychology, 67.* https://doi.org/10.1016/j.jenvp.2020.101388.

Luczak, A. (2020). *Deutschlands Energiewende – Fakten, Mythen und Irrsinn.* Springer.

Mackay, C. M., & Schmitt, M. T. (2019). Do people who feel connected to nature do more to protect it? A meta-analysis. *Journal of Environmental Psychology, 65*. https://doi.org/10.1016/j.jenvp.2019.101323.

Mackay, C. M., Schmitt, M. T., Lutz, A. E., & Mendel, J. (2021). Recent developments in the social identity approach to the psychology of climate change. *Current Opinion in Psychology, 42*. https://doi.org/10.1016/j.copsyc.2021.04.009.

Masson, T., & Fritsche, I. (2021). We need climate change mitigation and climate change mitigation needs the ‚We': A state-of-the-art review of social identity effects motivating climate change action. *Current Opinion in Behavioral Sciences, 42*, 89–96. https://doi.org/10.1016/j.cobeha.2021.04.006.

Mau, S., Lux, T., & Westheuser, L. (2023). *Triggerpunkte – Konsens und Konflikt in der Gegenwartsgesellschaft*. Suhrkamp.

McCright, A. M., & Dunlap, R. E. (2013). Bringing ideology in: The conservative white male effect on worry about environmental problems in the USA. *Journal of Risk Research, 16*(2), 211–226.

Milkau, U. (2022). *Operational resilience in finanzinstituten*. Springer Gabler.

Moore, J. W. (2020). Capitalocene & planetary justice. *Études digitales, 9*, 53–65. https://classiques-garnier.com/etudes-digitales-2020-1-n-9-capitalocene-et-plateformes-hommage-a-bernard-stiegler-capitalocene-planetary-justice.html. Accessed: Aug. 8, 2023.

Nachhaltigkeitsbarometer. (2023). Soziales Nachhaltigkeitsbarometer der Energie- und Verkehrswende 2023. https://ariadneprojekt.de/pressemitteilung/weitaus-mehr-menschen-als-angenommen-befuerworten-klimaschutz/. Accessed: Aug. 6, 2023.

NBC News. (2020). Quelle: https://www.nbcnews.com/politics/2020-election/newsom-gently-confronts-trump-climate-others-his-administration-aren-t-n1240095. Accessed: Aug. 12, 2023.

Pew Research Center. (2023). What the data says about Americans' views of climate change. https://www.pewresearch.org/short-reads/2023/04/18/for-earth-day-key-facts-about-americans-views-of-climate-change-and-renewable-energy/. Accessed: July 2, 2023.

Poortinga, W., Whitmarsh, L., Steg, L., Böhm, G., & Fisher, S. (2019). Climate change perceptions and their individual-level determinants: A cross-European analysis. *Global Environmental Change, 55*, 25–35.

Quent, M., Richter, C., & Salheiser, A. (2022). *Klimarassismus. Der Kampf der Rechten gegen die ökologische Wende*. Piper Verlag.

Rucht, D. (2023). Die Letzte Generation – Beschreibung und Kritik. *Ipb working paper*, 1/2023, Ipb.

Schaller, S., & Carius, A. (2019). *Convenient Truths. Mapping climate agendas of right-wing populist parties in Europe*. Adelphi consult GmbH.

Schulte, E., Scheller, F., Sloot, D., & Bruckner, T. (2022). A meta-analysis of residential PV adoption: The important role of perceived benefits, intentions and antecedents in solar energy acceptance. *Energy Research & Social Science, 84*, 102339.

Shaw, J. (2021). Feyerabend and manufactured disagreement: Reflections on expertise, consensus, and science policy. *Synthese, 198*(Suppl 25), 6053–6084.

Spiegel. (2023a). Thüringer CDU-Chef setzt AfD und Grüne als »Angstparteien« gleich. https://www.spiegel.de/politik/deutschland/thueringen-mario-voigt-setzt-afd-und-gruene-als-angstparteien-gleich-a-4e946555-82af-4ee2-8254-29af0f0605e4. Accessed: Aug. 7, 2023.

Spiegel. (2023b). Habeck macht »orientierungslose« CDU für AfD-Erfolge mitverantwortlich. https://www.spiegel.de/politik/deutschland/gruene-robert-habeck-macht-orientierungslose-cdu-fuer-afd-erfolge-verantwortlich-a-533573db-b113-4fb8-be45-8a42a4b17a0c. Accessed: Aug. 7, 2023.

Stanley, S. K., & Wilson, M. S. (2019). Meta-analysing the association between social dominance orientation, authoritarianism, and attitudes on the environment and climate change. *Journal of Environmental Psychology, 61,* 46–56.

Statista (2023). https://de.statista.com/statistik/daten/studie/1379717/umfrage/engagement-fuer-den-klimaschutz-in-deutschland-nach-parteien/. Accessed: Aug. 6, 2023.

Stollberg, J., & Jonas, E. (2021). Existential threat as a challenge for individual and collective engagement: Climate change and the motivation to act. *Current Opinion in Psychology, 42,* 145–150.

Tagesschau. (2023a). Ist das noch Wetter oder schon Klimawandel? https://www.tagesschau.de/wissen/klima/klima-wetter-100.html. Accessed: Aug. 6, 2023.

Tagesschau. (2023b). Merz relativiert Aussagen zu AfD-Zusammenarbeit. https://www.tagesschau.de/inland/innenpolitik/merz-kritik-104.html. Accessed: Aug. 7, 2023.

Tagesschau. (2023c). https://www.tagesschau.de/inland/union-beobachtung-letzte-generation-100.html. Accessed: Aug. 12, 2023.

Tagesschau. (2023d). Deutsche lehnen Straßenblockaden mehrheitlich ab. https://www.tagesschau.de/inland/innenpolitik/klima-proteste-umfrage-aktivisten-letzte-generation-100.html. Accessed: Aug. 12, 2023.

Tajfel, H., & Turner, J. C. (1979). An integrative theory of intergroup conflict. In W. G. Austin & S. Worchel (Eds.), *The social psychology of intergroup relations* (pp. 33–47). Brooks/Cole.

Tam, K. P., Chan, H. W., & Clayton, S. (2023). Climate change anxiety in China, India, Japan, and the United States. *Journal of Environmental Psychology, 87,* 101991.

Tol, R. S. (2023). The climate niche of homo sapiens. https://arxiv.org/abs/2306.00002. Accessed: 8. July 2023.

UNRIC – Regionales Informationszentrum der Vereinten Nationen. (2023). Juli 2023: Der heißeste Monat seit Aufzeichnungsbeginn; https://unric.org/de/juli-2023-der-heisseste-monat-seit-aufzeichnungsbeginn/. Accessed: Aug. 6, 2023.

Von Storch. (2018). https://twitter.com/Beatrix_vStorch/status/1024188890603835392?lang=de. Accessed: Aug. 7, 2023.

Von Storch (2023). https://twitter.com/Beatrix_vStorch/status/1685950749463674880. Accessed: Aug. 7, 2023.

Whitmarsh, L., Player, L., Jiongco, A., James, M., Williams, M., Marks, E., & Kennedy-Williams, P. (2022). Climate anxiety: What predicts it and how is it related to climate action? *Journal of Environmental Psychology, 83,* 101866.

Wolling, C., Becker, J., & Schumann (Eds.). (2023). *Klima(wandel)kommunikation: Im Spannungsfeld von Wissenschaft, Medien und öffentlicher Meinung.* Universitätsverlag, BoD–Books on Demand.

Xu, C., Kohler, T. A., Lenton, T. M., Svenning, J. C., & Scheffer, M. (2020). Future of the human climate niche. *Proceedings of the National Academy of Sciences, 117*(21), 11350–11355.

ZDF. (2023a). Hunderte Waldbrände in Kanada außer Kontrolle. ZDF, 8.07. 2023. https://www.zdf.de/nachrichten/panorama/waldbraende-kanada-hitze-umwelt-klima-100.html. Accessed: Aug. 12, 2023.

ZDF. (2023b). Juli wohl heißester Monat seit Jahrtausenden. https://www.zdf.de/nachrichten/panorama/juli-hitze-welt-rekord-klima-wandel-wetter-un-100.html. Accessed: Aug. 6, 2023.

ZDF. (2023c). AfD-Parteitag in Magdeburg. https://www.zdf.de/nachrichten/politik/afd-parteitag-wahlprogramm-europawahl-100.html. Accessed: Aug. 8, 2023.

ZDF. (2023d). Scholz glaubt nicht an dauerhafte AfD-Stärke. https://www.zdf.de/nachrichten/politik/kanzler-scholz-sommer-pressekonferenz-100.html. Accessed: Aug. 8, 2023.

War and Peace 18

In "Science as Art," Paul Feyerabend confesses that Grazia Borrini-Feyerabend, his wife and great love, brought him clarity about the great problems of our time: "The problem of peace in its various forms—peace with our fellow human beings, even when they hold different opinions; peace with other nations, even when this involves admitting great mistakes; and the problem of peace with nature, even if that means that we no longer regard nature as our slave, but as an equal partner in life" (Feyerabend, 1984, pp. 12).

18.1 Pacifism I—Looking Back

Bertha von Suttner (1843–1914), born Countess Kinsky von Wchinitz und Tettau, wrote as a journalist and pacifist against war and advocated for a peaceful and humane world. In 1889, her novel "Lay Down Your Arms" was published. "Today," it says optimistically, "there is hardly anyone left who does not dream this dream (of peace, WF) or who would not admit its beauty. And there are also sentinels—truly vigilant sentinels—who want to awaken humanity from the long sleep of barbarism and who gather together energetically and purposefully to raise the *white flag*. Their battle cry is: 'War on war'; their slogan—the only word that could still save Europe, which is arming itself for ruin—is: 'Lay down your arms!'" (von Suttner, 1892, p. 304; emphasis in the original 1889 edition).

On December 10, 1905, Bertha von Suttner was awarded the Nobel Peace Prize. This brought to a close a year that was by no means peaceful. At the beginning of January 1905, German troops captured one of the important Nama fortresses in what is now Namibia, then called "German South West Africa" by the

W. Frindte, *Against Narrow-Mindedness and Chauvinism—Journeying Through Absurd Times with Paul K. Feyerabend*,
https://doi.org/10.1007/978-3-658-50113-6_18

Germans, in a bloody battle. The genocide of the Herero and Nama had begun in 1904 and ended in 1908. Between 60,000 and 80,000 Herero and about 10,000 Nama lost their lives (Kößler & Melber, 2004; see also: Chap. 15.2). On January 22, 1905 (December 9 in the Julian calendar), 150,000 people in Russian St. Petersburg set out for the Tsar's residence. They wanted to peacefully petition him for humane working and living conditions, for religious tolerance, and for democratic conditions. Tsarist soldiers fired on the people. Between 150 and 300 people were killed and over 1000 wounded. The "Bloody Sunday of St. Petersburg" marked the beginning of the ultimately unsuccessful Russian February Revolution. In May 1905, Japanese ships defeated the Russian fleet, sealing Russia's defeat in the Russo-Japanese War. Also in May, there were repeated pogroms against the Jewish population in Russian-Ukrainian territories. In the city of Zhytomyr, 200 people died. In October 1905, according to official figures, more than 400 Jews were killed in a pogrom in Odessa incited by the Tsarist police. Unofficial sources spoke of over 1000 dead (Weinberg, 1987, p. 53). Meanwhile, the Allied powers Great Britain, France, and Russia and the Central Powers Germany and Austria-Hungary were preparing to "reorder" the world. In 1908, Austria-Hungary annexed the territories of Bosnia and Herzegovina, which until then had belonged to the Ottoman Empire. In April 1909, Ottoman troops committed a massacre in Adana, Turkey, in which 20,000 Armenians were murdered.[1] Then, from 1911 to 1912, Italy and the Ottoman Empire fought the Italo-Turkish War, which is said to have been one of the most brutal wars. It was about supremacy in the Mediterranean, North Africa, and Yemen. This was followed in 1912 and 1913 by the Balkan Wars, in which Bulgaria, Greece, Montenegro, and Serbia tried to break Ottoman dominance.

On November 6, 1911, in Budapest and on November 9 in Bucharest, Bertha von Suttner gave a lecture in which she addressed the Italo-Turkish War and the peace movement. She denounced what she saw as Italy's war of aggression against the Ottoman provinces of Tripolitania and Cyrenaica, areas in present-day Libya. This attack had been preceded by an Italian ultimatum demanding that the Ottoman government cede the provinces to Italy. After the Ottoman government rejected the demand, the Italian government declared war on the Ottomans. "The war," said Bertha von Suttner, "was proclaimed, so to speak, only as a defense against war, as a defense against attacks, as a last resort in self-defense—and

[1] Up to one million Armenians (if not more) fell victim to the organized genocide by the Turks in 1915 and 1916 (Schladebach, 2005, pp. 98).

then suddenly, through that Italian ultimatum, through the declaration of war and attack that followed after twenty-four hours, despite offers to negotiate, war reappeared in all its arrogant, brutal autocracy, without any disguise, without a mask—as the affirmation of the unrestrained right of the stronger, of the urge to conquer, of tyranny! What an example! What a precedent, which in the future will open the floodgates to violence!" (v. Suttner, 1911, p. 316). After Italy had violated the Hague Convention of 1907[2], the question arose as to whether pacifism was now bankrupt. No, says Bertha von Suttner. It is armed peace that is bankrupt, as are the official assurances and treaties behind which there was no sincerity. The Italian attack on Tripoli and the North African regions was an assault in which the European governments, because they behaved so passively, were equally involved. "The war itself, as it is now beginning to unfold in Tripoli, will confirm the arguments of the pacifists and show the world how right they are in claiming that there is no profitable war of conquest, that there are no more glorious, joyful, military parades. The world has changed" (v. Suttner, ibid., p. 318). Therefore, from now on, the fighters for peace will present their demands even more resolutely. "They will close ranks more tightly and new supporters will join them."

And they did their best, the peace activists, the pacifists. As early as 1892, Bertha von Suttner and Alfred Hermann Fried[3] founded the pacifist journal "Die Waffen nieder!" (Lay Down Your Arms!) and the "German Peace Society" (DFG). Shortly before the outbreak of the First World War, the DFG had 98 local chapters with a total of about 10,000 members (Dingeldey, 2015). Incidentally: At that time, the Social Democratic Party of Germany had about one million members (SPD Geschichtswerkstatt, 2023).

In 1910, the American steel magnate and philanthropist Andrew Carnegie founded the "Carnegie Endowment for International Peace" and endowed it with ten million US dollars (which would be about $3.5 billion today). In the same year, the British journalist Norman Angell published his bestseller "The Great Illusion. A Study of the Relation of Military Power in Nations to their Economic and Social Advantage," in which he argued that war was irrational because it

---

[2] The Hague Convention is based on the peace conferences in The Hague, Netherlands, and the humanitarian principles adopted there for armed conflicts. In 1907,—among other countries, e.g., Germany, France, Russia, and Austria-Hungary—Italy and the Ottoman Empire also signed the convention (The Hague, 2010).

[3] Alfred Hermann Fried (1864–1921) received the Nobel Peace Prize in 1911 together with the Dutch jurist and politician Tobias Asser.

would lead to major economic and social upheavals (Van den Dungen, 2014). In 1907, delegates at the Congress of the Second International in Stuttgart passed a resolution calling on workers to do everything possible to prevent a war. This was reaffirmed in 1910 at the Congress in Copenhagen and again in 1912 at the Congress in Basel, convened on the occasion of the Balkan War. However, the congresses also revealed differences between the Social Democrats from France and Germany. While the French demanded concrete measures against a possible war (in the event of war, workers in all countries were to go on general strike to paralyze the industries producing materials essential for armaments), the German Social Democrats around August Bebel (with the exception of Rosa Luxemburg, who had been active in the German Social Democracy since 1898) pressed for more cautious and general commitments (Musner, 2014, pp. 57).

Shortly before the outbreak of the First World War, numerous anti-war demonstrations took place in the European capitals. In Berlin, it is said that around 100,000 people demonstrated for peace in July 1914 (Dingeldey, 2015).

Pacifists and peace activists also faced rejection and anger. They were denounced as "enemies of the state" and "bad patriots." In February 1914, Rosa Luxemburg was charged with "disobedience to authority" because, at a rally in Frankfurt am Main in September 1913, she had called on people not to raise their weapons against their French and foreign brothers (see also: Piper, 2019, pp. 450).

But it was too late. The pacifists and peace activists could no longer prevent the First World War. On August 4, 1914, Bertha von Suttner had died six weeks earlier and August Bebel had already passed away on August 13, 1913, the SPD parliamentary group in the German Reichstag declared its approval of the war credits. Only a few SPD members of parliament voted against them, such as Rosa Luxemburg, Clara Zetkin, Karl Kautsky, and Franz Mehring.[4] "I will shoot myself in the head, that will be the best protest against the betrayal of the party and may yet bring the masses of workers to their senses," Rosa Luxemburg is said to have declared immediately after the approval of the war credits, according to Hugo Eberlein[5] (Utopie kreativ, 2005, p. 357).

[4] In December 1915, almost half of the SPD deputies in the Reichstag then voted against further credits.

[5] Hugo Eberlein (1887–1941), member of the SPD and friend of Rosa Luxemburg, was a founding member of the KPD and was sentenced to death and shot in 1941 during the Stalinist terror in the Soviet Union.

On August 1, 1914, the German Reich declared war on Russia, followed by a declaration of war on France on August 3. General mobilization began, and with a tremendous surge of nationalism and chauvinism, German soldiers marched off to war. On August 4, Kaiser Wilhelm II delivered his famous throne speech: “I no longer recognize parties, I only recognize Germans.” Whether the German population was actually as enthusiastic about the war as is often assumed, however, is open to doubt.

Between 8.5 and 17 million people, soldiers and civilians, are said to have died in the steel storms and trench warfare, from bombings, disease, famines, and ethnic cleansing during the First World War. Tens of thousands of soldiers died from poison gas attacks (Woyke, 2016, p. 70). The poison gas for the Germans was developed by the German chemist Fritz Haber (1868–1934). In his view, “using all means to defeat the enemy” was his “moral duty” and that of every German (quoted in Szöllösi-Janze, 1998 p. 261). In 1918, Haber received the Nobel Prize in Chemistry for the ammonia synthesis process he developed (the Haber–Bosch process). In 1933, Haber emigrated to England. He was Jewish.

And who was to blame for the war? The Jews, the left-wing parties, and the pacifists. Even during the war, accusations of “Jewish shirkers” and “Jewish war profiteers” surfaced (Ullrich, 2019, p. 58). After his fall, Kaiser Wilhelm II blamed the “Jewish” leaders of the labor movement for the defeat in the war and called for the “extermination” of the Jews. The myths of the “stab-in-the-back legend” and of the “pacifist, international, antimilitarist, and revolutionary subversion of the army” (as claimed by Prussian General Hermann von Kuhl, Field Marshal Paul von Hindenburg, and later Adolf Hitler) were invoked to shift responsibility for Germany’s defeat in the First World War onto political opponents (Lobenstein-Reichmann, 2002).

The opponents of the war and the pacifists, however, saw their struggle for a “nonviolent world” vindicated by the horrors of the Great War and the intensified militarism after the war’s end. Even during the war, for example, the *International Women’s League for Peace and Freedom* was founded. In England, the *Union of Democratic Control* was established, an association dedicated to maintaining international relations. In the USA, the *League to Enforce Peace* was established, which, though conservative in orientation, was committed to reshaping international relations and laid the groundwork for the later League of Nations (von Gerlach, 1924). On October 2, 1919, left-republican intellectuals, including Carl von Ossietzky, Kurt Tucholsky, the mathematician Emil Julius Gumbel (1891–1966), and the physician and sociologist Georg Friedrich

Nicolai (1874–1964)[6], founded the "Peace League of War Participants." In October 1920, the 9th Pacifist Congress took place in Braunschweig, attended by about 1500 people from several German pacifist organizations, which in 1922 merged to form the German Peace Cartel (DFK) (Braunschweig, 1920). The chairman of the DFK was Ludwig Quidde (1858–1941), who was awarded the Nobel Peace Prize in 1927 (Holl, 2002).[7] In 1929, shortly before its dissolution, the DFK represented a total of 28 German peace organizations with about 100,000 members. In The Hague, the *International Congress for a New Peace* was held in December 1922, organized by the International Women's League for Peace and Freedom (Nattermann, 2022). In the summer of 1926, the Jew, writer, pacifist, and gay rights activist Kurt Hiller founded the *Group of Revolutionary Pacifists*. In 1929, the group adopted a new program, which rejected not only wars of aggression but also wars of national defense and execution or sanction wars, and called for general disarmament and conscientious objection. The main source of wars was seen as the "capitalist social order"; therefore, the struggle for peace must also be "work for the social revolution" (Hiller, 1929, p. 173).

The list of pacifist activities in the 1920s is not exhaustive. Especially since pacifism in the Weimar Republic manifested itself in very different forms, structures, and contents, and its actors were by no means united.

Radical-religious pacifism, pacifism oriented toward international law, anarchist pacifism, socialist pacifism, and many other movements attempted to articulate themselves for peace and against war during the interwar period. Communists of Leninist persuasion also took a position on pacifism during this time. The line had been set by Vladimir Ilyich Lenin (1870–1924) himself. In March 1915, he addressed the delegates of the "Social Democratic Workers' Party of Russia" (SDAPR), who were meeting in Bern: "Pacifism and abstract preaching of peace are a form of misleading the working class" (Lenin, 1960a, p. 152; written on March 4, 1915). The idea that a so-called democratic peace could be achieved without revolutions was fundamentally wrong. This was also directed against Karl Kautsky and other social democratic "renegades." In August 1915, Lenin became even clearer and emphasized that, when it came to assessing a war, Marxists differed vastly from "pacifists" and "anarchists." "Marxism is not pacifism," Lenin wrote in 1915 in "Socialism and War" (Lenin, 1960b, p. 331; written

[6] Georg Friedrich Nicolai was known as Georg Friedrich Lewinstein until 1897, after which he took the surname of his great-granduncle Otto Nicolai (https://dewiki.de/Lexikon/Georg_Friedrich_Nicolai).

[7] The prize of the same name, awarded annually by the German Foundation for Peace Research, is named after Ludwig Quidde (German Foundation for Peace Research).

in July/August 1915). Peace-loving sentiments among the masses could indicate that the "reactionary character" of a war had been recognized. To exploit such a sentiment was the "duty of all Social Democrats." But: "Whoever wants a lasting and democratic peace must be for civil war against the governments and the bourgeoisie" (ibid., p. 317). In short: peace could not be achieved without the overthrow of capitalism and without revolution.

Against such a doctrine, as one might call it, the aforementioned Kurt Hiller took a stand. He quotes "[...] one of the best minds of Moscow," Nikolai Ivanovich Bukharin, who, in an "ABC of Communism" (German edition 1921), claimed in the above-mentioned Leninist sense that pacifism was the misleading and stupefaction of the working masses, because pacifism kept the masses from the armed struggle for communism. Hiller countered: "The 'goal' of the pacifist sermon is not: specifically the working class, but is: to divert all people and groups of people (primarily: the nations) 'from armed struggle'" (Hiller, 1923, p. 297).

Shortly before his escape to Switzerland, Ludwig Quidde expressed himself even more clearly: "The Soviet Union, the international homeland of pacifists! That is sheer madness! There can hardly be sharper opposites than Bolshevism and pacifism. Even the opposition between pacifism and fascism is not sharper. The highest of all principles of pacifism is the nonviolent resolution of all disputes, both nationally within states and internationally between peoples. Both fascism and Bolshevism are based on the principle of violent suppression of all dissenting tendencies. They are embodiments of the same principle of brutal violence, only with opposing tendencies toward socialist ideas" (Quidde, 1932, pp. 227). Ludwig Quidde must have known about the forced collectivization of agriculture decided by Stalin and the CPSU in 1929. The peasants were expropriated or forcibly resettled, to Siberia, the Urals, Kazakhstan. Those counted among the "counterrevolutionary activists" were either sent to labor and penal camps or executed. Between five and six million people were affected by this "war against the village" (Hildermeier, 2000, p. 595). Between 530,000 and 600,000 people lost their lives. The forced collectivization did not improve the supply situation. On the contrary, it led to major famines in 1932 and 1933, in which an estimated further six million peasants died. Quidde could certainly not have known about the famine catastrophe in Ukraine and the "Great Terror" in 1932, but he was aware of the contempt for humanity in the early days of the Soviet Union. Incidentally, Lenin's directives regarding peace, which could only be achieved through violent revolution, were not questioned by philosophers in the GDR either (see, e.g., Hörz, 1976; Kiessling & Scheler, 1976).

Far more important than the disputes between the pacifist movements are the reactions with which militarists, nationalists, fascists, and communists responded to the diverse efforts of pacifists and opponents of war. On January 15, 1919, Freikorps soldiers murdered Rosa Luxemburg and Karl Liebknecht. On February 21, 1919, the first Minister President of Bavaria, Kurt Eisner, was killed. Eisner had initially supported the war against Russia and then became a determined opponent of the war during World War I. In May 1919, Freikorps soldiers murdered the philosopher and pacifist Gustav Landauer. Matthias Erzberger, a leading politician of the Centre Party, was also killed by nationalist Freikorps soldiers in August 1921. Before World War I, Erzberger had fueled the militarists' war frenzy with chauvinistic tirades, but during the course of the war he too became a critic of German war policy and later advocated for a peace of understanding as well as for the establishment of the League of Nations as an important instrument for future world peace (Dowe, 2011). In June 1922, the murder of Walther Rathenau (1867–1922), then Foreign Minister of the Reich, followed. Rathenau was a Jew and a German. The murder had been commissioned by the "Organisation Consul," a terrorist brigade of the Freikorps. Rathenau was not a pacifist. Even in 1918, he called for the continuation of the war, if necessary through the mobilization of the masses. After the defeat, he advocated for the fulfillment of reparations to the victorious powers and saw in this a "path to peace" (Volkov, 2012, p. 219).

The swan songs for humanity and the protagonists of inhumanity soon became impossible to ignore or overlook. On January 30, 1933, Reich President Paul von Hindenburg appointed the chairman of the NSDAP, Adolf Hitler, as Reich Chancellor. After Hindenburg's death, Hitler declared himself the Führer of the Germans, and the terror began its course. In February 1933, the National Socialists set fire to the Berlin Reichstag. Subsequently, the KPD and soon after the SPD were banned, rivals were murdered in the "Röhm Putsch," and concentration camps for political opponents were established. On April 7, 1933, the National Socialists passed the "Law for the Restoration of the Professional Civil Service" (see also: Chap. 14). On August 6, 1933, the "German Peace Society" and the "German League for Human Rights"[8] were also classified as impermissible political

[8] "The roots of the International League for Human Rights go back to 1914. In that year, the *Bund Neues Vaterland* was founded, which advocated for international understanding and the immediate end of the German war of aggression. From 1922, the association called itself the *German League for Human Rights* to emphasize its increasing cooperation with the French League for Human Rights. Until its ban in March 1933, the *German League for Human Rights* was committed to securing the democratic rights enshrined in the Weimar Constitution. It warned against the growing militarism and fascism" (Internationale Liga für Menschenrechte, Ludwig Quidde Foundation).

societies (Bittner, 1987, p. 174). Kurt Tucholsky had already left Germany in 1929, Ludwig Quidde fled to Switzerland in 1933 at the age of 74. Emil Julius Gumbel lost his professorship at the University of Heidelberg, fled to France in 1933, and to the USA in 1940. Georg Friedrich Nicolai, who had vehemently opposed German war propaganda at the beginning of World War I, fled to Chile in 1933 and became a professor of physiology there. Others who did not manage to escape in time were arrested, such as Kurt Hiller, Erich Mühsam, or Carl von Ossietzky (1889–1938). Erich Mühsam, who, like Kurt Eisner, Gustav Landauer, and Ernst Toller, was one of the protagonists of the Munich Soviet Republic in 1919, advocated more anarcho-pacifist, anti-authoritarian positions (rejection of any rule and violence against people). Mühsam was murdered in 1934 in the Oranienburg concentration camp. Carl von Ossietzky died on May 4, 1938, as a result of his imprisonment in the Esterwegen concentration camp and the mistreatment he suffered there. He was unable to accept the Nobel Peace Prize awarded to him in 1936. After his imprisonment in several concentration camps, Kurt Hiller fled via Prague to London in 1934. There he founded the Freedom League of German Socialists. In 1955, he returned to Germany and founded the "Neo-Socialist League" to campaign against any war of aggression. (See the excellent biography of Daniel Münzner, 2015, on Kurt Hiller).

For the National Socialists, pacifists were among the enemies of the movement. Although Hitler appeared to speak positively about pacifism in "Mein Kampf," he made it clear that anyone who wished for the "victory of the pacifist idea" must "[…] use all means to strive for the conquest of the world by the Germans" (Hitler, A., "Mein Kampf." A Critical Edition, 2016, p. 304). By September 1, 1939, at the latest, it was clear that Hitler meant business. The Second World War began, marking the low point of humanist efforts. In 1940, the Wehrmacht invaded Denmark, Norway, later the Netherlands, Belgium, Luxembourg, and France. At the beginning of 1940, SS Reichsführer Himmler ordered the construction of the Auschwitz concentration camp. In mid-October 1940, more than 400,000 Jews were confined to the Warsaw Ghetto. Also in October 1940, Soviet troops occupied the Baltic States. On November 14, German bombers destroyed the city center of the English city of Coventry. On June 22, 1941, German troops invaded the Soviet Union. In December 1941, Germany declared war on the USA. With the Wannsee Conference in January 1942 and the subsequent mass deportations, the final phase of the National Socialist extermination policy began, culminating in the systematic, mass murder of the Jews. Six million Jews died in Auschwitz and Treblinka, in Belzec and Sobibor, in Majdanek and Chelmno.

Pacifism could not prevent it. Perhaps because pacifism, in its various shades, relied too much on the mediation of attitudes or calls for nonviolence, rather than on the revolt of peace-oriented movements.[9]

Paul Feyerabend, lest we forget him, was known to think little of the mere didactic transmission of values, norms, and convictions when it came to war and peace. In "Erkenntnis für freie Menschen", for example, one reads: "Not everything can be permitted. War enthusiasts, for example, should not be allowed to force peace-loving people into their war games. […] People can be educated so that they do not do certain things. They are taught 'humanity,' 'respect for life,' and other such nonsense, and it is then hoped that they will live according to these ideas. I consider this a childish optimism. No teaching or, rather, no *humanitarian* teaching that does not mentally castrate people achieves anything" (Feyerabend, 1980, pp. 295; emphasis in the original).

Certainly, this is once again a typically Feyerabendian provocation, whose challenge is lessened if one instead considers the practical alternatives that Feyerabend seems to prefer. In "Zeitverschwenung" (Killing Time), his autobiography, he recounts—as mentioned (Chap. 6)—the student unrest he experienced in Berkeley at the end of the 1960s: "During the so-called student revolution, I discussed the theories that had accompanied earlier revolutionary movements. Cohn-Bendit, Lenin's pamphlet *Left-Wing Communism: An Infantile Disorder,* essays by Chairman Mao, and Mills' *On Liberty* were on my reading list. I asked the students to organize discussions or prepare demonstrations instead of writing papers, and I encouraged outsiders to present their point of view. […] At my invitation, Vietnamese students explained the history of their country and the reasons for resistance. A group of young gay men described what it felt like to be a minority among nothing but narrow-minded and ignorant citizens" (Feyerabend, 1995, pp. 167, emphasis in the original). Feyerabend, however, was not entirely opposed to calls against war, crime, and murder. I have mentioned an example in Chap. 10 and will mention it again here. In the final essay of the book "Die Vernichtung der Vielfalt" (The Destruction of Diversity), Feyerabend engages with an appeal by philosophers (written, among others, by Gadamer, Derrida, Ricoeur, Rorty, Putnam). The philosophers had addressed the parliaments and

[9] Certainly, such revolts did occur. In January 1918, millions of people in Germany went on strike under the slogan "Peace and Bread" against the continuation of the war (Feldman et al., 1972). And during the Second World War, numerous groups organized resistance against war and National Socialism, such as the "Red Orchestra," the "Confessing Church" around Martin Niemöller, the "Kreisau Circle," and others (Benz, 2014).

governments of the world, calling for the promotion of philosophy education in schools. Feyerabend criticizes the fact that the real problems of our time are not even mentioned in the appeal, and asks: "What are the real problems? They consist of war, violence, hunger, disease, and environmental catastrophes. Warring parties have invented a wonderful instrument to 'resolve existing contradictions': ethnic cleansing. The appeal has nothing to say about these atrocities. [...] The philosophers and scientists who signed it would have done better to issue a harsh condemnation of the crimes and murders happening in our midst, together with an appeal to all governments to intervene and stop the killing, if necessary by military force. Such a condemnation and such an appeal would have been understood. It would have shown that philosophy cares about its fellow human beings" (Feyerabend, 2005, p. 295).

18.2 The "Cold War"—a Myth?

On May 8, 1945, the Second World War ended with Germany's defeat; on September 2, Japan capitulated. The atomic bombings of Hiroshima and Nagasaki on August 6 and 9, 1945, catapulted "[...] humanity into a new era, namely the atomic age [...]. However, this era can hardly last long: *Either we bring it to an end, or it will very likely bring about our own end*" (Chomsky, 2021, p. 20; emphasis in original). On October 24, 1945, the Charter of the *United Nations* came into force. World peace and international security were henceforth to be maintained, global problems solved through international cooperation and friendly relations among nations, and human rights promoted (The Charter). Three years later, on December 10, 1948, the Charter was expanded by the *Universal Declaration of Human Rights*. Just over two years earlier, on March 12, 1947, U.S. President Harry S. Truman, with the so-called *Truman Doctrine*, had dissolved the wartime coalition with the Soviet Union and declared that it must be U.S. policy "[...] to support free peoples who are resisting attempted subjugation by armed minorities or by outside pressures" (Truman, quoted in Woyke, 2016, p. 135). What was meant were the "free peoples" threatened by communism. This was directed against the Soviet Union and its attempts to expand its own spheres of power and ideological claims internationally. The immediate occasion for Truman's doctrine was the Soviet Union's attempt to support Greek anti-fascists and communists in their civil war against the conservative government. In response to Truman's doctrine, Soviet politician Andrei A. Zhdanov, a confidant of Stalin, announced the "Two-Camp Theory" on September 25, 1947. "Thus, *Zhdanov* also diagnosed a division of the world, an ideological and political split, with the main opponents being the USA and the Soviet Union, each leading its own 'camp'" (Woyke, ibid., p. 139; emphasis in original).

The "Cold War" began.[10] Strictly speaking, this war, which historians believe only ended in the autumn of 1989, was not cold at all. One can point to the so-called proxy wars as well as numerous seemingly regionally limited military conflicts to demonstrate the absurdity of the term.

Examples:
In the Korean War from 1950 to 1953, a military alliance led by the USA (following a decision by the UN Security Council, which the Soviet Union, however, boycotted) fought against North Korean and Chinese troops, who were equipped with weapons from the Soviet Union. Over four million people are said to have lost their lives in this war (Stöver, 2013, p. 108).

After U.S. President Eisenhower and other American politicians warned in 1954 about communist dictatorships in Latin America and Soviet influence there, the CIA supported a violent and illegal coup against the democratically elected president of Guatemala, Jacobo Árbenz Guzmán (see also: Epe, 2018).

From 1955 to 1975, war raged in Vietnam, Cambodia, and Laos. With support from the Soviet Union and China, North Vietnamese troops, together with the "National Front for the Liberation of South Vietnam" (the so-called "Vietcong"), fought against the soldiers of the "anti-communist" regime in South Vietnam. In violation of international law, the USA intervened in the war with ground troops, aircraft, and napalm, supporting the South Vietnamese regime. In April 1975, U.S. troops were forced to leave Vietnam defeated. Over 1.4 million people, including about 58,000 American and one million South Vietnamese soldiers, were killed in this war (see: Woyke, 2016, pp. 207). The numbers vary. The suffering is still felt today.

The civil war in Angola, which began in 1975 and still reverberates today, saw the "People's Movement for the Liberation of Angola" (MPLA) and the "National Front for the Liberation of Angola" (FNLA) fighting each other. Together with the "National Union for the Total Independence of Angola" (UNITA), they had initially, from 1961, directed their struggle against the Portuguese colonial power. Now it was about the claim to power within the country.

[10] The term was coined in 1946 by an aide to U.S. presidential advisor Bernard Baruch. It became widely known through the 1947 pamphlet "The Cold War. A Study in U.S. Foreign Policy" by Walter Lippmann, published in the New York Herald Tribune (Stöver, 2021).

The Soviet Union, East Germany, and Cuba supported the MPLA. The FNLA, which allied itself with UNITA, received financial, material, and military support from the USA and the South African apartheid regime; China also intervened on their side (see also: Reiber, 2002). 500,000 people are said to have died in the civil war.

The civil war in Syria, ongoing since 2011, has claimed over 350,000 civilian victims by 2023. They died from cluster bombs, rockets, chemical weapons, hunger, and lack of humanitarian aid. The dictatorial Syrian president Bashar al-Assad was supported, among others, by Russia, Iran, and China, while the USA, the EU, and Turkey side with the Syrian rebels. This war, too, is more than a civil war (Hammed, 2023).

After the communist-oriented "People's Democratic Party of Afghanistan" came to power in 1979 and pursued increased secularization of Afghanistan, a barely transparent civil war flared up. Leading politicians were assassinated, not least with the help of Soviet KGB agents. Amidst all this chaos, the Soviet Union under the leadership of Leonid Ilyich Brezhnev decided to invade Afghanistan on December 25, 1979. The resistance, which quickly intensified after the invasion and involved not only the regular Afghan army but especially religious fighters, the Mujahideen, was politically, financially, and militarily supported above all by the USA, Pakistan, and Saudi Arabia. The guerrilla tactics of the Mujahideen and the American Stinger missiles, with which the Mujahideen fought against Soviet attack helicopters, probably ultimately decided the war (Lüders, 2022). In February 1989, Soviet troops withdrew. One million dead were left behind. And the next civil wars soon followed, from which the Taliban emerged victorious. The stories are well known. "Operation Enduring Freedom" failed, as did the "defense of Germany at the Hindu Kush" (Jahn, 2012) or the peace agreement between the USA and the Taliban in 2020. On August 31, 2021, U.S. troops left Afghanistan almost in haste. The Bundeswehr also withdrew. At times, 100,000 U.S. soldiers and about 40,000 soldiers from allied NATO countries were stationed in Afghanistan. 3800 soldiers of the NATO alliance, including 59 members of the Bundeswehr, died there. According to estimates, over 39,000 civilian casualties are to be mourned (Goertz, 2022). Not to mention the wounded, persecuted, and those who fled.

The list of examples could go on. One could recall the absurdities of the Yugoslav wars in the 1990s, the genocides, massacres, and war crimes, mainly committed by the Serbs, the ambivalent role of Russia during the course of the war, and the NATO combat missions not covered by international law. The coup by the U.S.-backed military under General Pinochet against the democratically elected socialist president Salvador Allende in September 1973, the Chechen wars (1994–

1996; 1999–2009), the war in Yemen (since 2015), or the Gulf Wars (1980–1988; 1990–1991; 2003–2011) could also be analyzed in terms of their "proxy functions."

I will leave it at these enumerations. To speak and write of the "Cold War" seems to me to be too narrow-minded. Such a designation follows a "Western" semantics that ignores and seeks to erase the suffering of the numerous victims of the so-called proxy wars.

18.3 War or Special Operation

The *Heidelberg Institute for International Conflict Research* (HIIK) stated in the Conflict Barometer 2022, among other things: Over the course of the year, the HIIK was able to observe 363 conflicts worldwide. The number of "high-intensity wars" increased from 36 to 42 compared to the previous year (Conflict Barometer, 2022). These include, among others, the violent clashes between Armenia and Azerbaijan, wars in sub-Saharan Africa (including between Islamists and the governments of Nigeria, Cameroon, Chad, and Niger), limited wars in the Democratic Republic of Congo (Tutsi rebels against government troops), in Ethiopia (Tigray People's Liberation Forces against Ethiopian and Eritrean government troops), and Myanmar (opposition forces against government troops).

On February 24, 2022, Russia attacked Ukraine in order to "denazify" it. When the Russian invaders encountered fierce resistance from Ukrainians in September and October 2022, Russian President *Vladimir Vladimirovich Putin* threatened to defend his country with "all available means." In the "West," this was interpreted as a "nuclear threat." US President Joe Biden then stated that the risk of a nuclear catastrophe was higher than at any time since the Cuban Missile Crisis in 1962, and warned of a "nuclear Armageddon" (Tagesschau, 2022a). Meanwhile, Ukrainian President *Volodymyr Zelensky* called for making a Russian use of nuclear weapons impossible through a NATO preemptive strike, only to later qualify his statement (Tagesschau, 2022b). The Russian government or its spokesperson then condemned this as a "call to start World War III." And the Russian war of aggression against Ukraine continued (and continues). Russian missiles hit residential buildings, civilian infrastructure, marketplaces, hospitals. Russian soldiers are alleged to have committed massacres against the civilian population. The Russian government denies this.

On February 27, 2022, the German Chancellor declared that February 24 marked a turning point in the history of the European continent: "We are witnessing a turning point. And that means: The world after is no longer the same as the world before. At its core, it is about the question of whether might can break the law, whether we allow Putin to turn back the clock to the era of the great powers of the 19th century, or whether we have the strength to set limits for warmongers like Putin. That requires our own strength" (Scholz, 2022 p. 8). On April 27, 2022, the German Bundestag, by majority vote on a motion by the parliamentary groups of SPD, CDU/CSU, Alliance 90/The Greens, and FDP, approved comprehensive support and the delivery of heavy weapons to Ukraine. The AfD and The Left voted against it. The AfD argued that this would make Germany a party to the war. The Left feared further escalation and advocated for diplomatic efforts to end the war (German Bundestag, 2022).

However, the unity among the SPD, the Greens, the FDP, and the CDU did not last long. Over the following months (in fact, up to the present), the federal government and especially the Chancellor have repeatedly been accused of being too hesitant when it comes to the rapid delivery of weapons. The accusations have come and continue to come from the opposition, from leading politicians of the coalition parties, from EU countries, and especially from Ukraine itself. The accusations and disputes are well known and need not be recounted here.

Meanwhile, the NATO states—led by the USA—and the European Union are supporting Ukraine on a large scale in its defensive struggle with humanitarian, financial, and military aid. By early September 2023, the total financial volume pledged by EU countries to Ukraine amounted to 156 billion euros. 69.5 billion came from the USA and just under 37 billion from other countries (Source: IfW-Kiel, 2023). As of September 2023, Germany was the largest donor after the USA, with approximately 22 billion. In addition to political, financial, and humanitarian aid, Germany supplied extensive military equipment, such as armored combat vehicles like Gepard, Leopard, Marder, or Dachs tanks, air defense systems, anti-aircraft missiles, ammunition, special equipment, reconnaissance drones, etc. (Federal Government, 2023).

Destruction, flight, displacement, annihilation, and death were still part of everyday life for people in Ukraine even in autumn 2023. The figures for civilian and non-civilian casualties vary and are mostly unverifiable. The Heidelberg Institute for International Conflict Research relies on United Nations data and reports that in 2022 at least 6900 civilians were killed and at least 11,000 injured. Between 18,000 and 46,500 Ukrainian and between 30,000 and 60,000 Russian soldiers were killed in the war in 2022 (Conflict Barometer, 2022; p. 17). These figures, too, cannot be verified.

Many German politicians seem to agree that Ukraine is defending "[…] our freedom, our peace order as well, and we support them financially and militarily—as long as necessary. Period," as German Foreign Minister Annalena Baerbock said on August 28, 2022 (ZDF, 2022). Other politicians share this view, such as FDP politician Marie-Agnes Strack-Zimmermann, Green politician Anton Hofreiter, or CDU/CSU defense expert Roderich Kiesewetter. Is it the case, as it was in the Hindu Kush, that freedom and democracy in Europe are being defended in Ukraine, or is the courageous and painful defensive war of the Ukrainians first and foremost a defense of their own country?

The French philosopher, sociologist, and complexity theorist Edgar Morin, born in 1921, who fought as an officer in the French Resistance, put forward the thesis in February 2023 that there are three wars in Ukraine: "[…] the continuation of the internal war between the Ukrainian government and the separatist provinces, the Russian-Ukrainian war, and an internationalized anti-Russian political-economic war of the West, led by the USA" (Morin, 2023, p. 99).

This is a hard pill to swallow, which those who demand the unconditional victory of Ukraine over the Russian aggressor can hardly accept.

There is debate over whether, during the negotiations for German reunification in 1990, the USA promised the Soviet Union not to expand NATO into Eastern Europe, whether such a promise—if it existed—was broken by the "West," and whether NATO's eastward expansion ultimately exacerbated conflicts between the "West" and Russia. After extensive archival research, Joshua R. Itzkowitz Shifrinson concludes that the USA, under the leadership of George Bush Senior, misled then-Soviet President Mikhail Sergeyevich Gorbachev in order to expand US influence into Eastern Europe (Shifrinson, 2016). The historian and journalist Ignaz Lozo vehemently disputes this (Lozo, 2021).

That NATO expanded eastward is, however, a fact: Poland, the Czech Republic, and Hungary became NATO members in 1999, Latvia, Lithuania, Romania, Slovakia, and Slovenia in 2004, Croatia in 2009, Montenegro in 2017, and North Macedonia in 2020. However, the assumption that NATO's eastward expansion is the necessary and sufficient cause for the Russian invasion of Ukraine is speculative and absurd.

The war in Ukraine is not a "special operation." It is a war of aggression in violation of international law. Russia is the aggressor. Ukraine is defending itself against this. It needs political, financial, and military support from European countries, NATO, and the USA. However, to stylize this support as a "struggle of the free world" would be a relapse into the semantics of the Cold War.

In terms of international law, the criteria do not seem entirely clear when it comes to judging whether the "West," through its military assistance, has already become a party to the conflict and war. The *Scientific Services of the German Bundestag* stated in June 2023, among other things: "There are still no explicit legal opinions in the field of international law that generally classify the support of NATO states in favor of Ukraine as a form of participation in the conflict. However, a certain 'unease' about the legal and rhetorical 'orchestration' of Western support can hardly be concealed. The political catchphrase of the 'fighter jet *alliance*' in any case, even semantically, goes beyond the logistical process of delivering aircraft" (Scientific Services, 2023, p. 34; emphasis in the original).

The term "fighter jet alliance" was used by Ukrainian President *Volodymyr Zelensky* in May 2023, as he campaigned for the delivery of F-16 fighter jets to Ukraine in its fight against the Russian aggressor. In August 2023, Ukraine received commitments from the Netherlands and Denmark to provide such fighter jets. At the beginning of September 2023, Ukraine expanded its requests for military equipment. Ukrainian Foreign Minister *Dmytro Kuleba* called on Germany to deliver *Taurus* cruise missiles as quickly as possible. "Every additional day of discussions, coordination meetings, and reflection processes costs Ukrainian lives," argued Kuleba with strong moral force (ZDF, 2023). The line between legitimate military support for Ukraine and the status of being a party to the war seems to be becoming more permeable. In March 2023, leaked documents appeared in the media suggesting that 100 NATO special forces were on Ukrainian territory and training Ukrainian soldiers (t-online, 2023). Ukraine immediately denied this (Zeit Online, 2023).

Edgar Morin asks with concern: "Will the intensification of the international war within Ukraine spill over the country's borders, will the war spread to Europe and even beyond Europe?" (Morin, 2023, pp. 103). Is the war in Ukraine the continuation of a proxy war by other means?

18.4 Pacifism II—Today

"In a war," writes Paul Feyerabend, "a totalitarian state has a free hand. Humanitarian considerations do not restrict it. The only limitations are those of material, talent, people. A democracy, however, must treat an opponent humanely, *even if that reduces its chances of victory.* It is true—only a few democracies have ever lived up to this ideal, but those that have adhered to it have made an important contribution to our civilization" (Feyerabend, 1978, pp. 366; emphasis in the original).

It is quite possible that those who, during the war of aggression against Ukraine, were (and are) concerned about an escalation of the international situation have such a humane, perhaps even pacifist, ideal in mind. In April 2022, 28 German artists and intellectuals, initiated by Alice Schwarzer, drafted a letter to German Chancellor Olaf Scholz. Among the first signatories, alongside

Schwarzer, were filmmaker Andreas Dresen, actor Lars Eidinger, polymath Alexander Kluge, theologian and politician Antje Vollmer, Martin Walser, publicist Harald Welzer, and writer Juli Zeh (Emma, 2022). Within a few days, around 140,000 people signed this letter. The signatories appealed for no heavy weapons to be delivered to Ukraine and for efforts to be made as quickly as possible to achieve a ceasefire between Ukraine and Russia. A recapture of all Ukrainian territories occupied by Russia was deemed unrealistic. "The escalating armament taking place under pressure could be the beginning of a global arms spiral with catastrophic consequences, not least for global health and climate change. Despite all differences, the goal must be to strive for worldwide peace" (EMMA, 2022). Hardly had it been published when Alice Schwarzer and her fellow campaigners faced harsh criticism. In both traditional and social media, they were accused of arrogance, lack of empathy, and armchair pacifism (for a brief overview, see: Euronews, 2022). When Harald Welzer, as co-author of the open letter, tried to defend his position on the talk show *Anne Will* and warned of an "arms spiral," the then Ukrainian ambassador Andrij Melnyk, who was present, reacted angrily. It was easy, he said, to sit in a professor's office and philosophize. What Welzer was offering, he said, was morally depraved (Zeit Online, 2022).

At the time the open letter was published, opinion polls showed that the general population seemed divided in its views on the measures taken by the federal government in response to the Russian invasion. In a representative survey by the Forschungsgruppe Wahlen (Politbarometer, April 2022), 56% thought it was right to deliver heavy weapons to Ukraine, while 39% were against it. Criticism came from supporters of the Left Party (72%) and the AfD (56%). In January 2023, shortly after the federal government decided to deliver Leopard 2 tanks to Ukraine, 54% of respondents thought this was right, 38% were against it. There was majority approval in the western federal states (59% in favor, 33% against). In the eastern federal states, 35% were in favor and 57% against (Politbarometer, January 2023).

On April 29, 2022, Jürgen Habermas intervened in the debates about the "massive war of aggression in violation of international law" in the *Süddeutsche Zeitung*. On the one hand, he said he understood the "self-evident taking of sides against Putin." On the other hand, he was irritated by the self-assurance "with which, in Germany, the morally outraged accusers confront a reflective and restrained federal government." Is it not, Habermas asked, "a pious self-deception to bet on a

Ukrainian victory against the murderous Russian war machine without taking up arms oneself?" Besides, wars against a nuclear power cannot be won (Habermas, 2022, p. 12). He was surprised by the demands to bring Putin before an international court that is recognized neither by Russia and China nor by the USA. "Not that the war criminal Putin does not deserve to stand before such a court; but for now, he still holds a veto seat on the United Nations Security Council and can threaten his opponents with nuclear weapons. For now, an end to the war, at least a ceasefire, must be negotiated with him" (Habermas, ibid., p. 13). It may have been above all this last sentence that infuriated Habermas's opponents. Thus, Simon Strauss in the Frankfurter Allgemeine Zeitung accused him of "youth bashing" and "reckless denunciation of the Ukrainian government," and saw parallels between the arguments of Habermas and Alexander Gauland of the AfD (Strauss, 2022). Thomas Schmid of "Die Welt" considered it outrageous for Habermas to accuse the morally outraged of advocating the end of a mode of German politics based on dialogue and the preservation of peace (Schmid, 2022).

And the "shrill battle of opinions" between supporters and opponents of arms deliveries to Ukraine continued. In response to the open letter from Alice Schwarzer and her fellow campaigners, another group of intellectuals also published an open letter to the German Chancellor in May 2022. Among the 57 initial signatories were, among others, Gerhart Baum, Marieluise Beck, Maxim Biller, Marianne Birthler, Michel Friedman, Wolfgang Ischinger, Wladimir Kaminer, Daniel Kehlmann, Ilko-Sascha Kowalczuk, Sabine Leutheusser-Schnarrenberger, Igor Levit, Sascha Lobo, Eva Menasse, Herta Müller, and Marina Weisband. Among other things, they demanded that Ukraine be quickly equipped with all the weapons needed to repel the Russian invasion, that Russian energy exports be subjected to an embargo, and that Ukraine be offered a binding perspective for joining the EU. "German history commands every effort to prevent renewed wars of expulsion and annihilation. This applies all the more to a country in which the Wehrmacht and SS raged with utmost brutality" (Change.org., 2022). This is not only a strong moral argument, but also a rejection of pacifist pipe dreams. The Green politician and initial signatory of the second open letter, Marieluise Beck, had already bid farewell to pacifism during the Yugoslav wars. In May 2022, Claudia Roth, the Federal Government Commissioner for Culture and the Media, stated that the Greens had never been a pacifist party (Frankfurter Rundschau, 2022).

In February 2023, Sarah Wagenknecht and Alice Schwarzer launched a petition on change.org calling for an immediate halt to the escalation of arms deliveries to Ukraine. "Women have been raped, children terrified, an entire people traumatized. If the fighting continues like this, Ukraine will soon be a depopulated, destroyed country" (change.org., 2023). Therefore, the Chancellor should now take the lead, both at the German and European level, in forming a strong alliance for a ceasefire and for peace negotiations. Once again, criticism rained down. The demands of Wagenknecht and Schwarzer were described as naive, mendacious, immoral, and cynical. The political scientist Herfried Münkler, to name just one example, accused the authors and signatories of the petition of entering into a "complicity with the aggressor" (Kölner Stadt-Anzeiger, 2023).

Jürgen Habermas also weighed in once more. On February 14, 2023, he argued in the Süddeutsche Zeitung for negotiations with Russia. The then Ukrainian ambassador tweeted in response: "A disgrace for German philosophy. Immanuel Kant and Georg Friedrich Hegel would turn in their graves out of shame" (Welt Online, 2023).

Before one stubbornly rails against supposed opponents in the battle of opinions, one should first thoroughly read their arguments. Habermas does not call for a halt to arms deliveries to Ukraine, but rather considers a "face-saving compromise for both sides." He is concerned with a compromise for which the "West" must also feel responsible. And: Habermas explicitly writes of "[…] Ukraine in its courageous struggle against the illegal, indeed criminal, attack on the existence and independence of a sovereign state" (Habermas, 2023). As is well known, for Habermas, freedom and reason form a unity, albeit a dialectical one. "The rational freedom of the human will consists […] in the willingness to submit to the 'nature' of the moral law (… as an inclination toward the good) rather than to the nature of our affects and drives," as stated in Habermas's magnum opus from 2019 (Habermas, 2019, Vol. 2, p. 237). And on this foundation, inspired by Immanuel Kant, Habermas does not side with those who say we must overcome our fear of wanting to defeat Russia. Instead, he considers the statement "Ukraine must not lose the war" to be correct and advocates for negotiations.

While intellectuals in Germany—and not only there—argue about whether Ukraine should be supplied with heavy weapons and whether negotiations with Russia are possible, people are dying in Ukraine, but also in Russia, as a result of the war. In Germany—and not only there—people are taking to the streets to express their solidarity with Ukraine, to protest against the Russian war of aggression, and/or to demand immediate peace negotiations. These are by no means social movements with common ground. In many cases, there is no such overlap at all. The motives for participating in the demonstrations were and are very different. In May 2022 (N = 3045) and August 2022 (N = 2257), Gesine Höltmann and colleagues surveyed people who had participated in such demonstrations in Germany. Among other things, a polarization among the participants became apparent. 80% of participants in the peace demonstrations were critical or opposed to the Russian war of aggression. About 20% of respondents were described by the researchers as "pro-Russian" (Höltmann et al., 2022, p. 19).

On the anniversary of the Russian attack on February 24, 2023, at least 10,000 people gathered in front of the Russian Embassy in Berlin to express their solidarity with Ukraine. One day later, several thousand people again demonstrated in Berlin, demanding an end to arms deliveries to Ukraine. Around 13,000 people took part in the "Uprising for Peace" demonstration, which was initiated, among others, by Sarah Wagenknecht, including leading AfD politicians (rbb24, 2023). In Dresden, on February 24, several hundred people responded to a call from the AfD under the slogan "Create peace without weapons," the slogan of the old West German peace movement (MDR Sachsen, 2023).

At the Easter marches in April 2023, several thousand people demonstrated for peace in about 70 German cities, including Berlin, Bonn, Duisburg, Hanover, Leipzig, Munich, and Stuttgart. Participants carried placards and flags with the dove of peace, the rainbow symbol, and slogans such as "Negotiate instead of shooting," "Russia out of Ukraine," or "Peace with Russia." Flags of the *Alternative for Germany* were also spotted. The peace movements are divided.

Has pacifism failed, and has it become obsolete in the face of the Russian war of aggression? Tilman Brück, the former chairman of the Stockholm International Peace Research Institute (SIPRI), commented in a Deutschlandfunk podcast in March 2023: "The war in Ukraine is not proof of the failure of pacifism. On the contrary, the war in Ukraine is rather evidence of the fragility of a non-rule-based world order, an order based on the law of the strongest" (Brück, 2023).

Can pacifism prevent wars? History seems intent on teaching us otherwise. The Russell-Einstein Manifesto, initiated by Bertrand Russell and published on July 9, 1955, which, among other things, called for a reduction in nuclear arsenals, remained largely without consequence. The "Göttingen 18" group, a group of German nuclear scientists including Carl Friedrich von Weizsäcker, Otto Hahn, Werner Heisenberg, and Nobel laureate Max Born, was founded on April 12, 1957, to oppose the nuclear armament of the Bundeswehr. Although the Federal Republic did not become a nuclear power, the "Göttingen 18" could not prevent the stationing of US nuclear weapons on the territory of the old Federal Republic. The Easter marches, first launched in the old Federal Republic in 1960, or the Krefeld Appeal adopted by the West German peace movement in November 1980, could not avert the stationing of nuclear intermediate-range missiles in the old Federal Republic and in the GDR (Sonne, 2020).

But it is also true that "[…] there are more important things than winning wars…" (Feyerabend, 1978, p. 367).[11]

Let us give Tilman Brück the floor once again:

> "We must act pacifistically in the long term and with coherence. It is pointless to be a pacifist only once the shooting has already started. Of course, we can now correct mistakes of the past, such as reducing our dependence on fossil fuels from Russia, whose import finances the war in Ukraine. Unfortunately, we continue to make the same mistakes unperturbed in our trade with, for example, Saudi Arabia and China. […] In order to achieve pacifism, we must also regulate power in other dimensions, such as in the relations between the sexes, between humans and nature, between rich and poor, and between North and South. War is only one example of the abuse of power. #MeToo, racism, and the climate crisis are further examples, which often reinforce each other. […] It is not about utopia, revolution, or a grand gesture. Rules, norms, and institutions are continuously developed further. […] Pacifism cannot suddenly cure the mistakes of the past. But pacifism can help make the future safer, more equitable, and more sustainable" (Brück, 2023).

From the perspective of a pacifist, there is still much to be done. Once, the world was full of gods, according to Paul Feyerabend, 1988. From this, a drab material world emerged, which will hopefully transform into a more peaceful world, one in which matter and life, thought and feeling, innovation and tradition work together for the benefit of all (Feyerabend, 1988, p. 178).[12]

In this sense, pacifism is not merely the guilty conscience of those who, out of chauvinism and narrow-mindedness, abuse their power to widen the gap between rich and poor, between the sexes, between the Global South and the West, between cultures, nations, and social communities for reasons of profit. Pacifism, in all its shades, is a "humanism in action," to recall a phrase by Ernst Bloch (Bloch, 1985, p. 12), a humanism of "upright gait" in circumstances where freedom, equality, and solidarity still have to be fought for.

---

[11] The full quote is: "We must realize that there are more important things than winning wars, promoting science, finding the truth" (Feyerabend, 1978, p. 367). The reference to science and truth may at first seem puzzling in light of Feyerabend's critical stance toward science, but it is only part of his (scientific) pluralism. He was neither an enemy of science nor an opponent of truth-seekers, but a rebel who was averse to any glorification of science and truth. When asked what truth meant to him, he replied in his Trento lectures: "Well, sometimes one thing, sometimes another" (Feyerabend, 1998, p. 159; original: 1996).

[12] "It was once a world full of gods; it then became a drab material world and it will, hopefully, change further, into a more peaceful world where matter and life, thought and feelings, innovation and tradition collaborate for the benefit of all" (Feyerabend, 1988, p. 178).

18.5 Postscript: My Inconsistencies

> *"A uniform opinion may be the right thing for a church, for the intimidated or greedy victims of an (old or new) myth, or for the weak and compliant subjects of a tyrant"* (Feyerabend, 1986, p. 54; emphasis in the original).

It is also not always easy to have a uniform opinion on wars of aggression and wars of defense; especially not if, like me, one wishes to uphold the banner of pacifism. I do not want to be "treated as a scoundrel" (Bloch, 1985, p. 251) nor as a Putin apologist. In the Russian war of aggression, the aggressors, according to a report by the UN Commission of Inquiry on the Ukraine war (OHCHR, 2023), have committed numerous war crimes, crimes against humanity, and alleged genocide. Ukrainian children were deported, physical and moral coercion was used against civilians, people were unlawfully imprisoned and tortured. For this, the Russian aggressors must be held accountable. And Ukraine must be able to defend itself. The authors of the report also documented cases of war crimes by the Ukrainian armed forces (OHCHR, 2023, p. 16). These must also be prosecuted. Black and white are not the colors with which the war in Ukraine can be painted. There are more than 50 shades of gray that unsettle me in my pacifism.

On October 7, 2023, it was Shabbat and Jewish people were celebrating the festival of *Simchat Torah* (the festival of rejoicing in the Torah), when the Palestinian Hamas fired thousands of rockets at Israel. Militant Palestinians broke through the border fence between Gaza and Israel, attacking Israeli areas with motorcycles, quads, and paragliders. Shortly after 6:30 a.m., Hamas terrorists attacked an open-air festival in Kibbutz Re'im, near the Gaza Strip. The festivalgoers had come to celebrate a festival of "friendship, love, and freedom" at sunrise with international stars from the music scene. In Re'im, in the nearby Kibbutz Be'eri, in Kfar Aza, and other Israeli locations, the Palestinians committed unimaginable massacres. They slaughtered more than 260 people and burned their bodies; they beheaded children and babies and abducted over 240 festivalgoers as hostages to Gaza. According to figures released by the Israeli army after mid-October 2023, more than 1400 people were killed in Israel, and around 4000 were injured. Israel responded with airstrikes, in which at least 2000 people were said to have been killed in Gaza (ZDFheute, 2023). Analysts believe that in none of the wars Israel has had to fight were there more Israeli casualties in a single day than on this October 7, 2023. Eylon Levy, a former spokesperson for Israeli President Isaac Herzog, wrote on X that it was no exaggeration to say that October 7, 2023, was the darkest day in Jewish history since the end of the Holocaust (Tagesspiegel, 2023).

Hezbollah in Lebanon declared its solidarity with Hamas and stated that the time for revenge had come. And Iran, from which most of the weapons used by Hamas to attack Israel likely originated, threatened a "major earthquake" if Israel were to attack Hezbollah (Spiegel, 2023). Meanwhile, people in Germany gathered to express their solidarity with Israel. At pro-Palestinian demonstrations—also in Germany—not only were Palestinian flags waved and sweets distributed, but there were also threats of Israel's destruction.

Over the past three decades, I have visited Israel almost every year. My optimism for a peaceful solution to the "Middle East conflict" has become increasingly fragile (Frindte, 2020). Now, my pacifist notion of peaceful engagement with the Palestinian terrorists has also become obsolete. Those who kidnap people, humiliate them, who cold-bloodedly shoot families in their homes, who abuse frightened young women, who murder at a music festival, are not "freedom fighters." They are terrorists and criminals and must be punished, in the worst case, destroyed. Martin Buber and Franz Rosenzweig translated the sixth commandment (Lo tirzach) in the Decalogue not as "Do not kill," but as "Do not murder" (Ex. 20, 13) (Buber & Rosenzweig, 1987, p. 206). But that also means, in an emergency, you may kill. And the terror of the militant Palestinians is such an emergency.

References

Benz, W. (2014). *Der deutsche Widerstand gegen Hitler.* Beck.

Bittner, S. (1987). Das "Gesetz zur Wiederherstellung des Berufsbeamtentums" vom 7. April 1933 und seine Durchführung im Bereich der Höheren Schule. *Bildung und Erziehung, 40*(2), 167–182.

Bloch, E, (1985; Original: 1961). *Naturrecht und menschliche Würde.* Bloch Gesamtausgabe, Band 6. Suhrkamp.

Braunschweig. (1920). Eindrücke Vom IX. Deutschen Pazifistenkongress. *Die Friedens-Warte, 22*(7/8), 218–220.

Brück, T. (2023). Ist Pazifismus noch zeitgemäß? https://www.deutschlandfunkkultur.de/krieg-ukraine-pazifismus-noch-zeitgemaess-100.html. Accessed: Sept. 10, 2023.

Buber, M., & Rosenzweig, F. (1987). *Die Schrift. Band 1: Die fünf Bücher der Weisung* (Verdeutschung). Schneider.

Bundesregierung. (2023). Liste der militärischen Unterstützungsleistungen. https://www.bundesregierung.de/breg-de/schwerpunkte/krieg-in-der-ukraine/lieferungen-ukraine-2054514. Accessed: Sept. 9, 2023.

Change.org. (2022). Die Sache der Ukraine ist auch unsere Sache! https://www.change.org/p/die-sache-der-ukraine-ist-auch-unsere-sache?utm_source=share_petition&utm_medium=custom_url&recruited_by_id=835568b0-caa5-11ec-a137-77c2dc6ca625. Accessed: Sept. 10, 2023.

Change.org. (2023). Manifest für Frieden. https://www.change.org/p/manifest-f%C3%BCr-frieden. Accessed: Sept. 10, 2023.

Chomsky, N. (2021). *Rebellion oder Untergang! Ein Aufruf zu globalem Ungehorsam zur Rettung unserer Zivilisation.* Westend.

Conflict Barometer. (2022). https://hiik.de/conflict-barometer/current-version/?lang=en. Accessed: Sept. 9, 2023.

Der Spiegel. (2023). Iran droht Israel bei Angriff der Hisbollah mit »großem Erdbeben«. https://www.spiegel.de/ausland/israel-iran-droht-bei-angriff-der-hisbollah-mit-grossem-erdbeben-a-0ce35bbc-6fb4-4248-8fc5-5bc255fd1e98. Accessed: Oct. 14, 2023.

Deutsche Stiftung Friedensforschung. Ludwig Quidde-Stiftung. https://bundesstiftung-friedensforschung.de/ludwig-quidde-stiftung/. Accessed: Aug. 30, 2023.

Deutscher Bundestag. (2022). Antrag der Fraktionen SPD, CDU/CSU, BÜNDNIS 90/DIE GRÜNEN und FDP "Frieden und Freiheit in Europa verteidigen—Umfassende Unterstützung für die Ukraine". https://www.bundestag.de/dokumente/textarchiv/2022/kw17-de-selbstverteidigung-ukraine-891272. Accessed: Sept. 9, 2023.

Die Charta der Vereinten Nationen. UNRIC—Regionales Informationszentrum der Vereinten Nationen. https://unric.org/de/charta/. Accessed: Aug. 20, 2023.

Dingeldey, P. (2015). Bekämpfer des Krieges. Frankfurter Allgemeine. https://www.faz.net/aktuell/politik/der-erste-weltkrieg/pazifismus-vor-dem-ersten-weltkrieg-1914-13011786.html. Accesse: Aug. 20, 2023.

Dowe, C. (2011). *Matthias Erzberger: Ein Leben für die Demokratie.* Kohlhammer Verlag.

Emma. (2022). Der offene Brief an Kanzler Scholz. https://www.emma.de/artikel/offener-brief-bundeskanzler-scholz-339463. Accessed: Sept. 9, 2023.

Epe, M. (2018). Der guatemaltekische Bürgerkrieg im Schatten des Ost-West-Konflikts. In Derselbe, *Das Konzept des inneren Feindes in Guatemala. Politik in Afrika, Asien und Lateinamerika* (pp. 185–242). Springer VS.

Euronews. (2022). Empörung über offenen Brief an Scholz—Warnung vor dem 3. Weltkrieg. https://de.euronews.com/2022/04/29/emporung-uber-offenen-brief-an-scholz-warnung-vor-dem-3-weltkrieg. Accessed: Sept. 10, 2023.

Feldman, G.D., Kolb, E., & Rürup, R. (1972). Die Massenbewegungen der Arbeiterschaft in Deutschland am Ende des Ersten Weltkrieges (1917–1920). *Politische Vierteljahresschrift, 13*(1), 84–105.

Feyerabend, P.K. (1978). *Der wissenschaftstheoretische Realismus und die Autorität der Wissenschaften.* Vieweg & Sohn.

Feyerabend, P.K. (1980). *Erkenntnis für freie Menschen.* Suhrkamp.

Feyerabend, P.K. (1984). *Wissenschaft als Kunst.* Suhrkamp.

Feyerabend, P.K. (1986). *Wider den Methodenzwang.* Suhrkamp.

Feyerabend, P.K. (1988). Knowledge and the Role of Theories. *Philosophy of the Social Sciences, 18*(2), 157–178.

Feyerabend, P.K. (1995). *Zeitverschwendung.* Suhrkamp.

Feyerabend, P.K. (1998; Original: 1996). *Widerstreit und Harmonie. Trentiner Vorlesungen.* Passagen Verlag.

Feyerabend, P.K. (2005). *Die Vernichtung der Vielfalt. Ein Bericht,* herausgegeben von Peter Engelmann. Passagen Verlag.

Frankfurter Rundschau. (2022). Claudia Roth im Interview. https://www.fr.de/politik/claudia-roth-im-interview-die-gruenen-waren-nie-eine-pazifistische-partei-91575880.html. Accessed: Sept. 10, 2023.

Frindte, W. (2020). Aus meinem israelisches Tagebuch. https://www.hagalil.com/2020/05/israelisches-tagebuch/. Accessed: Okt.14, 2023.

Goertz, S. (2022). Afghanistan und die Taliban. *Ein Überblick. Essentials.* Springer VS.

Haag, D. (2010). *Die Haager Landkriegsordnung.* Europäischer Hochschulverlag.

Habermas, J. (2019). *Auch eine Geschichte der Philosophie: Band 1: Die okzidentale Konstellation von Glauben und Wissen. Band 2: Vernünftige Freiheit. Spuren des Diskurses über Glauben und Wissen.* Suhrkamp.

Habermas, J. (2022). Krieg und Empörung. *Süddeutsche Zeitung,* 29. April 2022, pp. 12–13.

Habermas, J. (2023). Ein Plädoyer für Verhandlungen. https://www.sueddeutsche.de/projekte/artikel/kultur/juergen-habermas-ukraine-sz-verhandlungen-e159105/?reduced=true. Accessed: Sept. 10, 2023.

Hammed, Y. (2023). Syrien zwischen Flucht und Krieg. In C. J. Henrich (Eds.), *Politik und Gesellschaft im Mittleren Osten: Eine Region im Spannungsfeld politischer und gesellschaftlicher Transformation* (pp. 209–223). Springer.

Hildermeier, M. (2000). Stalinismus und Terror. *Osteuropa, 50*(6), 593–605.

Hiller, K. (1923). Pazifismus und Kommunismus. *Die Friedens-Warte,* 23(9/10), 296–299.

Hiller, K. (1929). Das neue Programm der Revolutionären Pazifisten. *Die Friedens-Warte, 29*(6), 172–174.

Hitler, A. (2016). *Mein Kampf.* Eine kritische Edition. Institut für Zeitgeschichte, Online-Ausgabe. https://www.mein-kampf-edition.de. Accessed: Apr. 12, 2023.

Holl, K. (2002). Der Historiker und Pazifist Ludwig Quidde(1858–1941). Träger des Friedensnobelpreises von 1927. *Die Friedens-Warte, 77*(4), 437–454.

Höltmann, G., Hutter, S., & Rößler-Prokhorenko, C. (2022). Solidarität und Protest in der Zeitenwende. *Reaktionen der Zivilgesellschaft auf den Ukraine-Krieg.* WZB Discussion Paper, No. ZZ 2022-601. Wissenschaftszentrum Berlin für Sozialforschung (WZB).

Hörz, H. (1976). Naturwissenschaft–Frieden–Verantwortung. *Deutsche Zeitschrift für Philosophie, 24*(1), 50–57.

IfW-Kiel. (2023). Ukraine Support Tracker: Europa sagt jetzt doppelt so viel Unterstützung zu wie die USA. https://www.ifw-kiel.de/de/publikationen/aktuelles/ukraine-support-tracker-europa-sagt-jetzt-doppelt-so-viel-unterstuetzung-zu-wie-die-usa/. Accessed: Sept. 9, 2023.

Internationale Liga für Menschenrechte. https://ilmr.de/menschenrechte. Accessed: Aug. 20, 2023.

Jahn, E. (2012). Die "Verteidigung Deutschlands am Hindukusch". Die deutsche Rolle in Afghanistan. In *Derselbe, Politische Streitfragen.* VS Verlag für Sozialwissenschaften.

Kiessling, G., & Scheler, W. (1976). Friedenskampf und politisch-moralische Wertung des Krieges. *Deutsche Zeitschrift für Philosophie, 24*(1), 37–49.

Kölner Stadt-Anzeiger (2023). "Gewissenloses Manifest" von Schwarzer und Wagenknecht. https://www.ksta.de/politik/herfried-muenkler-verlogenes-manifest-von-alice-schwarzer-und-sahra-wagenknecht-zur-ukraine-454517. Accessed: Sept. 10, 2023.

Kößler, R. & Melber, H. (2004). Völkermord und Gedenken. Der Genozid an den Herero und Nama in Deutsch-Südwestafrika 1904–1908. In M. Brumlik & I. Wojak (Hrsg.),

Völkermord und Kriegsverbrechen in der ersten Hälfte des 20. Jahrhunderts (S. 37–76). Campus.

Lenin, W. I. (1960a; geschrieben März 1915). Die Konferenz der Auslandssektionen der SDAPR. In W.I. Lenin (Ed.), *Werke, Band 21*. Dietz.

Lenin, W. I. (1960b; geschrieben Juli/August 1915). Sozialismus und Krieg. In W.I. Lenin (Ed.), *Werke, Band 21*. Dietz.

Lobenstein-Reichmann, A. (2002). Die Dolchstoßlegende. *Zur Konstruktion eines sprachlichen Mythos. Muttersprache, 112*(1), 25–42.

Lozo, I. (2021). *Gorbatschow. Der Weltveränderer.* Wissenschaftliche Buchgesellschaft.

Lüders, M. (2022). *Hybris am Hindukusch: Wie der Westen in Afghanistan scheiterte*. Beck.

MDR Sachsen. (2023). AfD und Pegida demonstrieren Hand in Hand in Dresden. https://www.mdr.de/nachrichten/sachsen/dresden/dresden-radebeul/ukraine-jahrestag-krieg-afd-pegida-100.html. Accessed: Sept. 10, 2023.

Mein Kampf. (2016). Eine kritische Edition. Institut für Zeitgeschichte München, Berlin. https://www.mein-kampf-edition.de/?page=Pref-Book%2Fstart.html. Accessed 30. Aug. 2023.

Morin, E. (2023). *Von Krieg zu Krieg*. Verlag Turia + Kant.

Münzner, D. (2015). *Kurt Hiller: Der Intellektuelle als Außenseiter.* Wallstein Verlag.

Musner, L. (2014). Waren alle nur Schlafwandler? Die österreichische Sozialdemokratie und der Ausbruch des Ersten Weltkrieges. In M. Mesner, R. Kriechbaumer, M. Maier, & H. Wohnout (Eds.), *Parteien und Gesellschaften im Ersten Weltkrieg* (pp. 55–69). Böhlau.

Nattermann, R. (2022). Internationaler Feminismus und Humanitarismus nach dem Ersten Weltkrieg. Akteurinnen der Women's International League for Peace and Freedom im Spannungsfeld von Internationalismus, Nationalismus und faschistischer Herrschaft in Europa. Europäische Geschichte. https://www.europa.clio-online.de/essay/id/fdae-112845. Accessed: Aug. 30, 2023.

OHCHR. (2023). Report of the Independent International Commission of Inquiry on Ukraine. https://www.ohchr.org/sites/default/files/documents/hrbodies/hrcouncil/coiukraine/A_HRC_52_62_AUV_EN.pdf. Accessed: Oct. 14, 2023.

Piper, E. (2019). *Rosa Luxemburg. Ein Leben*. Blessing.

Politbarometer. (April 2022). https://www.forschungsgruppe.de/Umfragen/Politbarometer/Archiv/Politbarometer_2022/April_II_2022/. Accessed: Sept. 10, 2023.

Politbarometer. (January 2023). https://www.forschungsgruppe.de/Umfragen/Politbarometer/Archiv/Politbarometer_2023/Januar_II_2023/. Accessed: Sept. 10, 2023.

Quidde, L. (1932). Ein Kampfkongress gegen den imperialistischen Krieg. *Die Friedens-Warte, 32*(8), 227–230.

RBB24. (2023). Rund 13,000 Menschen fordern Ende der Waffenlieferungen an die Ukraine. https://www.rbb24.de/panorama/beitrag/2023/02/wagenknecht-schwarzer-berlin-frieden-fuer-die-ukraine.html. Accessed: Sept. 10, 2023.

Reiber, T. (2002). Angola: Endlich auf dem Weg zum Frieden? *Die Friedens-Warte, 77*(3), 313–328.

Schladebach, M. (2005). Der türkische Völkermord an den Armeniern: Aktuelle Fragen aus europäischer Perspektive. *Südosteuropa. Zeitschrift für Politik und Gesellschaft, 01*, 96–108.

Schmid, T. (2022). Wo Jürgen Habermas irrt—und wo er richtig liegt. https://www.welt.de/kultur/plus238466335/Habermas-zum-Ukraine-Ueberfall-Wo-der-Philosoph-irrt-und-wo-er-recht-hat.html. Accessed: Sept. 10, 2023.

Scholz, O. (2022). Regierungserklärung von Bundeskanzler Olaf Scholz am 27. Februar 2022. https://www.bundesregierung.de/breg-de/suche/regierungserklaerung-von-bundeskanzler-olaf-scholz-am-27-februar-2022-2008356. Accessed: Sept. 9, 2023.

Shifrinson, J.R.I. (2016). Deal or no deal? The end of the Cold War and the US offer to limit NATO expansion. *International Security, 40*(4), 7–44.

Sonne, W. (2020). Massenproteste wie nie—und doch umsonst. In W. Sonne (Ed.), *Leben mit der Bombe: Atomwaffen in Deutschland* (pp. 305–316). Springer.

SPD Geschichtswerkstatt. (2023). https://www.spd-geschichtswerkstatt.de/wiki/1913. Accessed: Aug. 20, 2023.

Stöver, B. (2013). *Geschichte des Koreakrieges: Schlachtfeld der Supermächte und ungelöster Konflikt.* Beck.

Stöver, B. (2021). Kalter Krieg. In M.G. Festl (Ed.), *Handbuch Liberalismus* (pp. 441–448). Metzler.

Strauss, S. (2022). Habermas zum Ukrainekrieg: Sollen wir Putin um Erlaubnis fragen? https://www.faz.net/aktuell/feuilleton/debatten/juergen-habermas-aeussert-sich-zum-ukraine-krieg-17993997.html. Accessed: Sept. 10, 2023.

Szöllösi-Janze, M. (1998). *Fritz Haber, 1868–1934: Eine Biographie.* Beck.

t-online. (2023). Kämpfen Nato-Spezialkräfte in der Ukraine? https://www.t-online.de/nachrichten/ausland/internationale-politik/id_100158742/pentagon-leaks-nato-truppen-koennten-in-der-ukraine-kaempfen.html. Accessed: Sept. 10, 2023.

Tagesschau. (2022a). Biden warnt vor Nuklear-„Armageddon". https://www.tagesschau.de/ausland/amerika/atomwaffen-biden-usa-101.html. Accessed: Sept. 9, 2023.

Tagesschau. (2022b). Selenskyj relativiert "Präventivschlag"-Aussage. https://www.tagesschau.de/newsticker/liveblog-ukraine-freitag-185.html#Praeventivschlag. Accessed: Sept. 9, 2023.

Tagesspiegel. (2023). "Tödlichster Angriff auf Juden seit dem Holocaust". https://www.tagesspiegel.de/internationales/todlichster-angriff-auf-juden-seit-dem-holocaust-diplomaten-und-analysten-sind-von-hamas-attacke-entsetzt-10593685.html. Accessed: Okt. 14, 2023.

Ullrich, A. (2019). *Von "jüdischem Optimismus" und "unausbleiblicher Enttäuschung".* De Gruyter.

Utopie kreativ. (2005). Hugo Eberlein. Erinnerungen an Rosa Luxemburg bei Kriegsausbruch 1914, *Utopia kreativ, 174,* 355–362.

Van den Dungen, P. (2014). Pazifismus vor 1914. Wissenschaft und Frieden, Issue 3, pp. 46–49; https://wissenschaft-und-frieden.de/artikel/pazifismus-vor-1914/. Accessed: Aug. 20, 2023.

Volkov, S. (2012). *Walther Rathenaus—Ein jüdisches Leben in Deutschland.* Beck.

Von Gerlach, H. (1924). Der Pazifismus seit Kriegsende. *Die Friedens-Warte, 24*(1/3), 40–43.

Von Suttner, B. (1892). *Die Waffen nieder! Eine Lebensgeschichte.* Zwei Bände, Band 2. Edgar Pierson Verlag.

Von Suttner, B. (1911). Tripolis und die Friedensbewegung. *Die Friedens-Warte, 13*(11), 316–318.

Weinberg, R. (1987). Workers, pogroms, and the 1905 revolution in Odessa. *The Russian Review, 46*(1), 53–75.

Welt Online. (2023). Habermas plädiert für schnelle Verhandlungen mit Putin. https://www.welt.de/newsticker/dpa_nt/infoline_nt/Politik__Inland_/article243776579/Habermas-plaediert-fuer-schnelle-Verhandlungen-mit-Putin.html. Accessed: Sept. 10, 2023.

Wissenschaftliche Dienste des Deutschen Bundestages. (2023). Militärische Unterstützung der Ukraine: Wann wird ein Staat zur Konfliktpartei? WD 2-3000-023/23.

Woyke, W. (2016). *Weltpolitik im Wandel.* Springer VS.

ZDF. (2022). Baerbock garantiert Hilfen: "Die Ukraine verteidigt auch unsere Freiheit". https://www.zdf.de/nachrichten/politik/ukraine-baerbock-unterstuetzung-hilfe-krieg-100.html. Accessed: Sept. 9, 2023.

ZDF. (2023). Kiew bittet um neue Waffen. https://www.zdf.de/nachrichten/politik/taurus-marschflugkoerper-waffenlieferung-ukraine-krieg-russland-100.html. Accessed: Sept. 10, 2023.

ZDFheute. (2023). Angriffe von Hamas auf Israel. https://www.zdf.de/nachrichten/politik/israel-raketen-angriff-gazastreifen-palaestinenser-100.html. Accessed: Okt. 14, 2023.

Zeit Online. (2022). Harald Welzer kritisiert Botschafter Melnyk. https://www.zeit.de/kultur/2022-05/harald-welzer-melnyk-ukraine-krieg. Accessed: Sept. 10, 2023.

Zeit Online. (2023). Ukraine dementiert Präsenz von Nato-Kräften, Russland verstärkt Abwehr. https://www.zeit.de/politik/ausland/2023-04/ukraine-ueberblick-geleakte-dokumente-nato-saporischschja-russland. Accessed: Sept. 10, 2023.

Part VI
Conclusion?

"[...] yes and now, unfortunately, I have ordered a taxi, I must go now—good night, bye, bye..." (Feyerabend, 1980, p. 300).

From May 4 to 8, 1992, Paul Feyerabend gave lectures as a guest professor at the *Università degli Studi di Trento*. After the lectures, audience members had the opportunity to ask him questions. In response to one of the comments from the audience, Feyerabend replied: "[…] first, many scientists lie—not directly and shamelessly, for they are full of good intentions, but in an indirect way that is not even transparent to themselves. Second, of course, they will tell a story. They will tell what they did, what their results were, why they were not satisfied with them, and so on. In the end, a streamlined story will enter the annals of science. It will tell of facts and traditions" (Feyerabend 1998, p. 158). So I can only hope that the stories I tell in this book have not become too streamlined.

References

Feyerabend, P. K. (1980). *Erkenntnis für freie Menschen*. Suhrkamp.

Feyerabend, P.K. (1998, Original: 1996). *Widerstreit und Harmonie*. Trentiner Vorlesungen. Passagen.

19 Conflict and Harmony

> "Do human actions have effects on the world? Of course they do" (Feyerabend, 1998, p. 83).

The first publication of the Trento Lectures appeared posthumously in 1996 at the Laterza & Figli publishing house in Rome. An English version was published in 2011 under the sensationalist title "The Tyranny of Science," edited by Eric Oberheim (Feyerabend, 2011). The publication is based on transcriptions of the lectures, which Feyerabend himself had revised. In a review of the English edition, Daniel B. Kuby writes that he cannot promise the eager readers of Feyerabend's works any new significant insights should they turn to "The Tyranny of Science" (Kuby, 2014, p. 371). No, what Paul Feyerabend tells his audience is not new compared to his earlier accounts. Some of his stories are systematically elaborated in the also posthumously published book manuscript "Die Vernichtung der Vielfalt" (Feyerabend, 2005), on which he worked until his death (see also Chap. 10).

What is impressive, however, is how he tells his stories, how he moves from current events to physics, reviews Greek philosophers, tragedies, and comedies, discusses mathematics through Galileo, Descartes, Leibniz, Euler, and Newton, and brings in US President Clinton or the microbiologist and Nobel laureate Jacques Monod.

"Strictly speaking," Feyerabend writes in the German edition, "my lectures will be fairy tales that I spin around some events that are vaguely historical. That does not really trouble me, for I suspect that real historians also tell fairy tales, only that their tales are longer and more complicated—which does not mean that they could not be very interesting" (Feyerabend, 1998, pp. 23). There he is again,

W. Frindte, *Against Narrow-Mindedness and Chauvinism—Journeying Through Absurd Times with Paul K. Feyerabend*,
https://doi.org/10.1007/978-3-658-50113-6_19

the provocateur, who, however, at no point tries to persuade his audience that science is tyrannical. Instead, he wants to draw attention to the intricacy and complexity of the sciences, about which it is a mistake to speak in the singular (Kidd, 2011).

For example, he takes the Big Bang theory, the street battles between people of color and the police in Los Angeles in April 1992, and the wars in Yugoslavia at the time as occasions to ask whether there is such a thing as a harmonious unity of the world that science can recognize and clarify with a modern worldview. "In any case, the idea of a unified world to which we all belong is only one idea among many. It cannot be the standard for all the others. But even if the world were unified, it is by no means clear that a worldview would be the best means of finding one's way in it. Worldviews are not only incomplete, they are also deceptive, and, to use a somewhat pompous expression, they diminish our humanity" (Feyerabend, 1998, p. 21).

And yet people seem eager to construct a unified, contradiction-free picture of the world. The search for the laws that hold the world together at its core is part of these efforts. Paul Feyerabend shows in his lecture journey through the history of philosophy and rationalism how this endeavor has taken shape, how, for example, the ancient philosophers such as Thales, Xenophanes, Plato, or Aristotle, or contemporary scientists—whom Feyerabend sometimes calls "odd birds" (ibid., p. 39)—tried to pin down the world through their abstractions. At the same time, over time and under certain social conditions, people have developed practical knowledge as craftsmen, engineers, doctors, or ingenious experimenters, for example, and this "[…] practical knowledge goes into detail and far surpasses what modern people, including scientists, can offer in this respect" (Feyerabend ibid., p. 92). Between theoretical and practical knowledge, there seems—if one follows Feyerabend—to have long been an agreement or harmony in history, which today has given way to conflict. And yet this practical knowledge is there and "[…] *it should be used*." Scientists and philosophers are not necessarily closer to reality than other people, than magicians, acupuncturists, birdwatchers, cooks, engineers, or people from the neighborhood (ibid., p. 126).

How, for example, can an ecological system be understood? Well, by working together with the people who have lived in the region for generations and who have the knowledge and can tell the stories about themselves and the region. And so Feyerabend once again—just as he did many years earlier—advocates for close "[…] cooperation between experts and the people whose environment the experts want to assess, change, and improve" (Feyerabend, 1998, p. 69). Such an approach is not only successfully practiced in many countries; crucially, "[…] this approach is much more humane than the purely objective procedure, which does not treat ordinary people as friends or potential collaborators, but as not always welcome, because rather disruptive, sets of variables" (ibid., p. 70).

Feyerabend thus offers a pluralistic, emancipatory, and humanistic approach, not to simply resolve the conflict between scientists and laypeople, between theorists and practitioners, or to force it into a long-outdated pseudo-harmony, but to transform it into forms in which it—the conflict—can move. This offer is by no means as self-evident as it may appear or as is expressed in critical reviews of the "Trento Lectures" (e.g., Sankey, 2012, p. 476). Feyerabend neither privileges practice (as everyday practice or experimental practice) over theorizing, nor does he express aversions to scientific abstractions. He does, to be sure, criticize in his usual manner the "ideology" of those scientists who claim that only scientific methods are the royal road to knowledge. Certainly, he sympathizes with practice, which seems more democratic to him (Feyerabend, 1998, p. 124), but he considers the relationship between theory and practice to be a complicated matter and science to be a craft that connects theory and practice in a dialectical way (Kuby, 2014, p. 374). In this respect, and this is precisely the core of Feyerabend's offer, one can (must?) also criticize science without becoming a scientist oneself. A critique of science is appropriate in dealing with the findings of psychology (Chap. 14), cancel culture (Chap. 15), crises, catastrophes, and wars (Chap. 16), and much else. "A *democratic critique of science* is not only not an absurdity—it *belongs to the nature of knowledge*" (Feyerabend, 1998, p. 53; emphasis in the original).

What does that mean? I venture an interpretation of the lessons I have received from Paul Feyerabend: This critique is—in most cases—a conflict in which people (scientific experts and laypeople) encounter each other who practice different language games, to recall Wittgenstein. Two different language games (or "phrase-regimes," as Jean-François Lyotard says in "The Differend"; Lyotard, 1989; original: "Le Différend", 1983), such as scientific and everyday bodies of knowledge, habits, traditions, and myths, cannot be judged and evaluated either from the perspective of the other language game and its rules (or criteria) or from the perspective of an overarching ideology or worldview (such as Enlightenment or rationalism).

That social groups or communities nevertheless repeatedly attempt to do this, trying to make their own particular ideas, values, and ideologies the standard of judgment, is something Paul Feyerabend has denounced in many of his publications. Different social groups or interpretive communities can argue about their bodies of knowledge, habits, traditions, and myths; they can argue about whether and to what extent the bodies of knowledge, habits, traditions, and myths in question construct reality, and thus to what extent their enactments are coherent; consensus in such disputes will not be achieved. The respective language games of the disputants can only be accepted, that is, tolerated, as real—because they exist—but different (each one different for the other) possibilities for speaking about and living out human forms of life and knowledge.

It is like a *dialectical contradiction,* or better: it is a dialectical contradiction that opens up and develops when two or more social groups (or interpretive communities) encounter each other within a *shared* living space, but attempt to shape it with different conventions, traditions, and myths. The points of friction in the conflict between the various cultural systems (between experts and laypeople) are found at the boundaries and zones of conflict where the unknown and unfamiliar collide with the known and conventional. Knowledge must be measured not only by whether it can be made practically effective, but also by the extent to which it relates to unfamiliar, seemingly erroneous, or absurd spheres of our realities; that is, by the extent to which, through dispute, the unknown and the new can be brought forth instead of the familiar. This means that the question is to what extent, through the conflict of social language communities, through the diversity of their bodies of knowledge, traditions, conventions, and myths, and through the dispute over this diversity, *new* questions and perspectives for engaging with the world can be raised.

"Anything goes: that's true! The most surprising things lead to great discoveries! Those who believe that new things can be found by sticking strictly to well-marked paths are mistaken. One cannot predict which crazy move will lead to a new insight or a new discovery. The move is only 'crazy' when compared to the prevailing opinion of the time in which one lives" (Feyerabend, 1998, p. 163).

That would then also be the lesson I have drawn from the works of Paul Feyerabend: A science that does not transcend its boundaries, that permanently accepts the framework set by its myth, its meta- and epistemology, as well as various social factors, I call *cold science.* A science of this kind floats above everything and all too quickly loses its connection to the lifeworld. It becomes an end in itself. *Hot science,* on the other hand, thrives on plurality, change, creativity, its processual character and its humanitarian aspirations, on self-reflection at various levels. I feel committed to the ideal of a hot science.

References

Feyerabend, P. K. (1998, Original: 1996). *Widerstreit und Harmonie.* Trentiner Vorlesungen. Passagen Verlag.

Feyerabend, P. K. (2005). *Die Vernichtung der Vielfalt. Ein Bericht,* herausgegeben von Peter Engelmann. Passagen Verlag.

Feyerabend, P. K. (2011). *The tyranny of science.* Polity.

Kidd, I. J. (2011). Paul Feyerabend – The Tyranny of Science, ed. Eric Oberheim. Polity. *The British Journal for the History of Science, 44*(4), 576–577.

Kuby, D. (2014). Review of Paul Feyerabend „The Tyranny of Science“, edited by Eric Oberheim. Polity Press. https://philarchive.org/archive/KUBROP. Accessed: Sept. 14, 2023.

Lyotard, J.-F. (1989, original: 1983). *Der Widerstreit.* Wilhelm Fink Verlag.

Sankey, H. (2012). Philosophical fairytales from Feyerabend, Review of Paul Feyerabend „The Tyranny of Science“. *Metascience, 21,* 471–476.

"Good Night, Bye, Bye..." 20

A few weeks before his death on February 11, 1994, Paul Feyerabend writes: "I do not want to die now, after having found my balance, also in my private life. I would like to stay with Grazia, support her and encourage her when the work becomes exhausting. [...] I want that after my death, not essays and not final philosophical explanations of mine remain, but love. I hope that it will endure and not be too much affected by the manner of my passing, without a death struggle, without leaving behind bad memories. Whatever happens now, our little family can live forever—Grazia, myself, and our love. That is what I wish for: not that my mind lives on, but only the love" (Feyerabend, 1995, pp. 248).

Is dying not something absurd? "Of course it is! We live in an absurd world!" (Feyerabend, 1992, p. 137).

References

Feyerabend, P. K. (1992; Original: 1989). *Über Erkenntnis. Zwei Dialoge.* Campus.

Feyerabend, P. K. (1995). *Zeitverschwendung.* Suhrkamp.

W. Frindte, *Against Narrow-Mindedness and Chauvinism—Journeying Through Absurd Times with Paul K. Feyerabend*, https://doi.org/10.1007/978-3-658-50113-6_20

MIX
Papier aus verantwortungsvollen Quellen
Paper from responsible sources
FSC® C105338

If you have any concerns about our products,
you can contact us on
ProductSafety@springernature.com

In case Publisher is established outside the EU,
the EU authorized representative is:
Springer Nature Customer Service Center GmbH
Europaplatz 3, 69115 Heidelberg, Germany

Printed by Libri Plureos GmbH
in Hamburg, Germany